THE WORLD
世界建筑事务所精粹
ARCHITECTURAL II
FIRM SELECTION

OFFICE BUILDING / COMMERCIAL BUILDING / CULTURAL
BUILDING / RESIDENTIAL BUILDING / COMPLEX BUILDING
办公建筑 / 商业建筑 / 文化建筑 / 住宅建筑 / 综合建筑

深圳市博远空间文化发展有限公司 编

天津大学出版社
TIANJIN UNIVERSITY PRESS

PREFACE

PREFACE

In the past few decades, the practice of architectural design has developed towards the postmodern design concept with collage, mixture and coexistence from the modernism on the basis of functionalism and single typology. Rationality turns to be the theme of the development clue of architecture from the modernism stage dominated by the geometric form to the postmodernism stage with multiple mixing symbols. Both the architecture and the social activities of human penetrate and promote each other. The history, economics, philosophy, sociology, etc. have exerted unprecedented influence on the theory and practice of architecture design, which enable the contemporary buildings to obtain the ecological characteristics with multi-elements. The thoughts of architecture such as the phenomenology, semiology, structuralism and deconstruction endow the architecture in this era with symbolic interpretations and create the works for this era.

The innovative and pioneering architects open up new possibilities of the functions, spaces and forms of the architecture with non-traditional innovative spirit and creative practices. The design methodologies based on the information technology, such as the nonlinear design, parametric design and virtual construction, greatly enrich the design measures of the architects, and open the gate towards the architecture in the new era. The unrealized design creations such as non-standardization, structural skin, free form and surreal characteristics under traditional design conditions have displayed themselves with the development of the digital technology, information technology, structure and material technology. The architecture breaks through the restraints from the structure and function, and the shape and space tend to be more complicated and intriguing, displaying the unique characteristics and times imprinting of the contemporary architecture. Extensive excellent cases and architectural practices are provided in this book, which offers architectural cases with different functions, scales, fields and cultural backgrounds. Readers can spy into the influences and expressions of the above mentioned architectural thoughts and design methodology in the excellent architectural creations, thus achieving thought-provoking inspiration and apperception.

Wang He
January 28, 2013

序言

在过去的几十年，建筑设计的实践从功能主义和单一类型学基础上的现代主义向后现代的拼贴、混合、共存并置的设计思想发展。建筑从强调几何形体构成的现代主义时期，到多元符号混合的后现代主义时期，理性成为这一发展线索的主题。建筑与人类的社会活动相互渗透、相互促进。历史学、经济学、哲学、社会学等都对建筑设计理论与实践产生了空前的影响，使当代建筑呈现出多元并存的生态特征；现象学、符号学、结构主义、解构主义等建筑思潮都给这个时代的建筑赋予了标志性的注释，并创造出属于这个时代的作品。

勇于创新和开拓的建筑师以颠覆传统的创新精神和创作实践，开拓了建筑功能、空间、形式的新可能。非线性设计、参数化设计、虚拟建造等后工业时代以信息技术为基础的设计方法论极大地丰富了建筑师的设计手段，打开了通向新时代建筑的大门。非标准化、结构性表皮、自由形体、超现实的特征等在传统设计条件下难以实现的设计创作，随着数字技术、信息技术、结构与材料技术的发展，一一呈现在世人面前，建筑自身突破了结构与功能的约束，造型与空间更趋于复杂与耐人寻味，表现出当代建筑独有的特征和时代印记。

本书提供了大量的优秀范例和建筑实践，呈现了不同功能、不同规模、不同地域和文化背景的建筑案例。读者可以从中窥探到以上建筑思潮和设计方法论在优秀建筑创作中的影响与展现，并从中得到思考性的启发与感悟。

王禾

2013 年 1 月 28 日

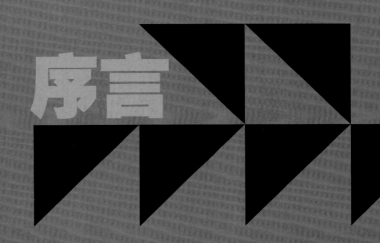

目 录
CONTENTS

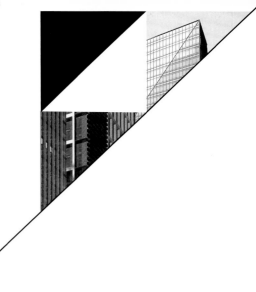

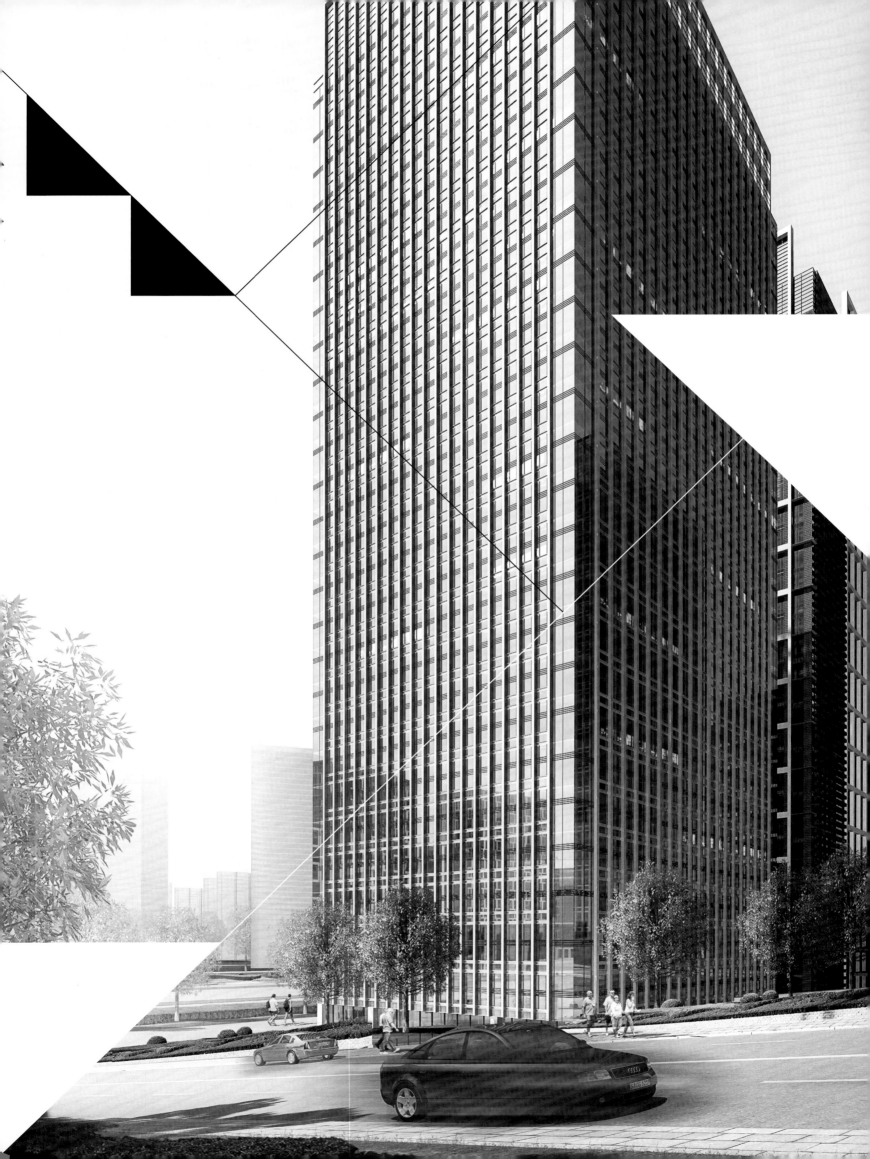

办公建筑　OFFICE BUILDING

Office building nowadays occupies a pretty high amount only second to residential house. Since we have entered a wholly new era of information due to the endless stream of emerging high-tech in the 21st century. The development of office building, as one of the main places to collect and handle information, is beyond comparison. According to an estimate, over a half of our population will work in office buildings in the middle of this century. So from this point of view, 21st century is "the century of office building".

As a kind of place to collect, handle and generate various administrative, scientific and commercial information, office building belongs to the infrastructure construction for social reproduction. It is a kind of place where information is produced, as well a place where we live. So the comfort level of the office directly affects the working efficiency of the staff. That is the reason why the requirement of a humanized design of a modern office building is exalted. Hence it is demanded that the design of the inner space of an office building should be as favorable to staff's working mind and behavior as possible. Similarly, a balance between the elegant, cozy and humanized outer and inner spatial environment is required. All these will greatly enhance the physical and psychological development of the staff, as well as the working efficiency.

Meanwhile, the spatial construction of office building is varying to adapt new official functions and new requirements for a higher comfort level. It is not only a request for a reasonable schedule of traffic routes, but also a result of spatial organization with a general consideration of elements such as aesthetics and technique. A mature and independent development system of office building construction is now gradually taking shape. Architects of office building are continuously seeking for brand-new design methods and architectural styles to capture the spirit of the era, and to create office works with the most era features and modern science and techniques.

It is gradually an important premise of architectural design to develop sustainably, environment-friendly and energy-savingly. How to maximize the function without the negligence of saving energy, and to make it a goal to be environment-protecting are among the key considerations of the architects. At present, the status of office building in the city is getting more and more important as its quality and quantity are increasing. It has become a significant part of the skyline of the city. Some momentous office buildings are even considered as landmarks of the city.

　　目前办公建筑已成为除住宅外，数量最多的一类建筑。在高新技术层出不穷的21世纪，人们进入了崭新的信息时代，办公建筑作为收集和处理信息的主要场所之一，其在信息时代的发展与过去已不可同日而语。据预测，到21世纪中叶，将有一半以上的人口在办公建筑里工作。从某种角度来讲，21世纪可谓是"办公楼的世纪"。

　　作为收集、处理和产生各种行政、科研、商务信息的场所，办公建筑是社会再生产的基础性建筑。办公室既是信息生产的场所，又是人们生活的场地，其环境的舒适度直接影响着员工的办公效率，因此，当代办公建筑设计对办公场所的人性化要求也逐渐提高，这就要求办公建筑内部空间的环境向着最有利于人员办公心理和行为的方向发展，同时要兼顾内外空间环境的美观、舒适，充满人性关怀，这些对员工身心健康和工作效率的提高都能起到重要的促进作用。

　　同时，办公建筑的空间构成也在不断适应新的办公功能和舒适性要求的变化，这不仅是合理安排交通路线的要求，也是综合考虑审美、技术等多种因素后的空间组合的结果。办公建筑已逐渐形成了成熟而独立的发展体系，建筑师在不断寻求新的设计方法、建筑形式，以捕捉时代精神，创造出最具有时代特色，能够显示现代科技的办公建筑作品。

　　可持续发展和绿色环保理念逐渐成为建筑设计的重要前提，如何在实现功能最大化的同时实现节约能源、绿色环保的目标，成为建筑师在设计办公建筑时考虑的重点之一。如今，随着办公建筑数量的增加和品质的提升，它在城市中的地位也愈加重要，成为城市天际线的重要组成部分，尤其是一些重要的办公建筑，已作为城市地标而成为城市的象征。

办公建筑　OFFICE BUILDING

WUHAN CENTER
武汉中心

Architects: ECADI
Location: Wuhan, China
Site area: 28,100 m²
Gross floor area: 355,760 m²

设计机构：华东建筑设计研究院
项目地点：中国武汉市
总用地面积：28 100 平方米
总建筑面积：355 760 平方米

The project is located in the southwest corner of central business district of Wuhan, facing the central square on the northeast and east, the hotel site on the west, the water park on the south. The project is the first and the most important component to implement the development strategy in CBD core area.
Wuhan Center tower is located in the middle site near to the intersection of the city roads. The podium and city square are arranged at the north and east in tower preposed and podium postposed pattern. The tower preposition can maximally use the both sides of the city road to give convenient traffic flows and pedestrian flows. The tower as a distinct landmark is directly toward the Mengze Lake. The commercial space encloses the Citizen Square by podium postposition to lead a direct contact with city public space. The large podium space is integrated with underground hub to improve the efficiency of the development. As for the form, Wuhan Center is like a sailing boat with full strengths and desires, marching forward courageously in the economic waves.

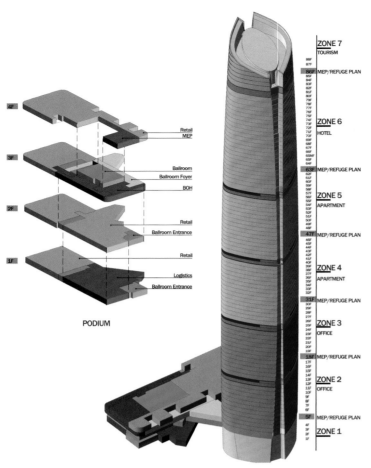

ZONE 7
TOURISM

88F
87F
8GF MEP/REFUGE PLAN
85F
84F
83F
82F
81F
80F
79F
78F
77F
76F ZONE 6
75F HOTEL
74F
73F
72F
71F
70F
6SF
66F
65F
64F
6GF MEP/REFUGE PLAN
61F
60F
59F
58F
57F ZONE 5
56F APARTMENT
55F
54F
53F
52F
51F
50F
49F
48F
47F MEP/REFUGE PLAN
45F
44F
43F
42F
41F
40F
39F ZONE 4
38F APARTMENT
37F
36F
35F
34F
33F
32F
31F MEP/REFUGE PLAN
29F
28F
27F
26F
25F ZONE 3
24F OFFICE
23F
22F
21F
20F
19F
18F MEP/REFUGE PLAN
17F
16F
15F
14F
13F ZONE 2
12F OFFICE
11F
10F
9F
8F
7F
6F
5F MEP/REFUGE PLAN
4F
3F
2F ZONE 1
1F

Retail
MEP

Ballroom
Ballroom Foyer
BOH

Retail
Ballroom Entrance

Retail

Logistics
Ballroom Entrance

PODIUM

Analysis 1 分析图 1

项目位于武汉中心商务区西南角,其东北面、东面为中心区广场,西面为旅馆用地,南面为水景公园。该项目是实施 CBD 核心区开发战略的第一个项目,也是最重要的一个项目。

武汉中心塔楼位于用地中部,靠近城市道路交叉口,裙房和城市广场布置在用地北侧和东侧,采用塔楼前置、裙楼后置的模式。塔楼前置可以最大化利用基地两边的城市道路,便于组织机动车流和人流;塔楼直接朝向梦泽湖,标志性强烈。裙楼后置使商业空间围合市民广场,与城市公共空间直接联系;裙楼大空间与地下枢纽站空间整合,提高开发效率。在造型上,武汉中心状若帆船,寓意满载希望与力量,在经济的浪潮中乘风破浪、勇往直前。

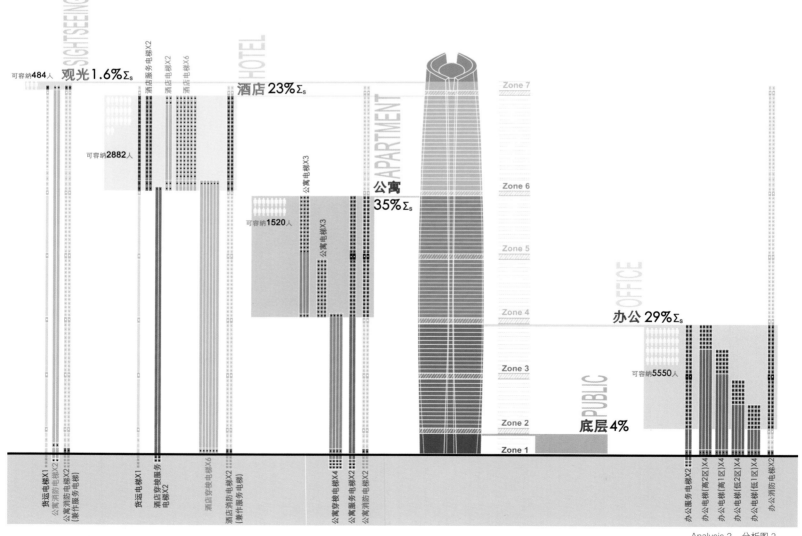

Analysis 2 分析图 2

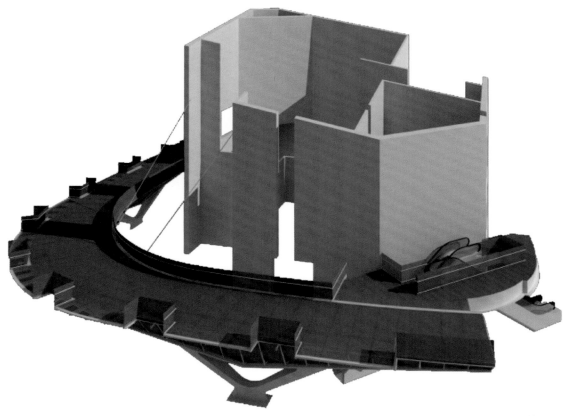

Modelling 1　模型图 1

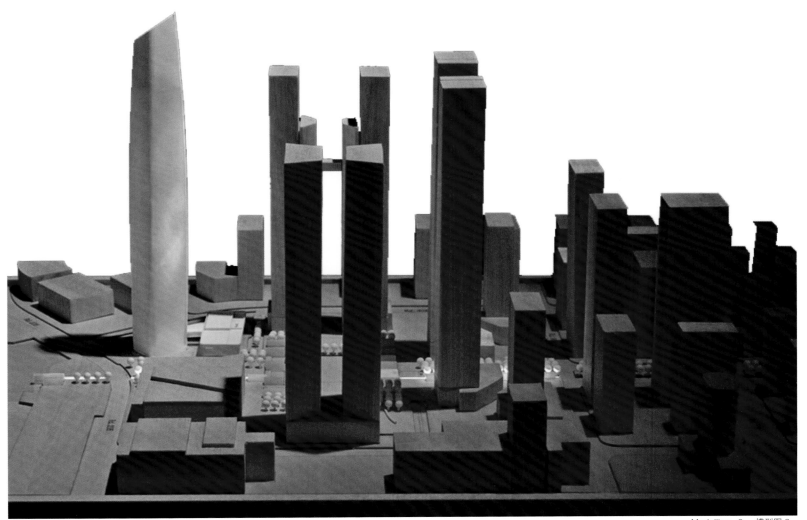

Modelling 2　模型图 2

EOS — MICROSOFT EUROPE HEADQUARTERS OFFICE COMPLEX

EOS——微软欧洲总部办公大楼

Architects: Arquitectonica
Associate architects: Agence d'architecture Bridot Willerval
Client: Bouygues Immobilier
Location: Issy-les-Moulineaux, France
Gross floor area: 46,000 m²
Height: 42 m
Photographer: Eric Morrill, Paul Maurer

设计机构：Arquitectonica
合作设计单位：Agence d'architecture Bridot Willerval
客户：Bouygues Immobilier
项目地点：法国 Issy-les-Moulineaux
总建筑面积：46 000 平方米
高度：42 米
摄影：Eric Morrill, Paul Mamurer

Project Description or Summary
This project is designed for the new European headquarters of Microsoft. The Bouygues Immobilier and the Generali Group are developing this new office complex of 46,000 m² on a spectacular site on the banks of the Seine River in a Paris suburb of Issy les Moulineaux. The development consists of three buildings, each 15,000 m², which are connected by courtyards, retail and 3 levels of parking.

Design Intent
The site development is defined by three volumetric typologies.
The Bar: A rectangular prism is placed along the SCNF track, establishing an urban edge and anchoring the development. This rectangle relates geometrically to the orthogonal texture of the neighborhoods beyond. It also provides a buffer from the SCNF activity, through its glassy transparency, also visually connects the project to the Issy District.
The Towers: Three sculptured volumes stand forward of the rectangular bar. They reach out towards the river and appear as freestanding objects in the composition. Two of them intersect the bar, one is free. These forms are intended to be evocative and character defining. In their dynamism they imply movement, in contrast with the static nature of the bar. Shallow curved glass surfaces begin apart at the bar and gently converge to a point as they reach towards the quay. The resultant space between them opens up vistas towards the river and exposes the buildings with equal presence to the roadway and the quay.
The Base: An organic, undulating form constitutes the podium from which the buildings rise. This new geography or a man-made topography conceals the numerous functions of the building base and provides an unexpected setting for the composition. The podium blends with the property's grounds along the quay, with the adjacent park, and with the new proposed park between the office development and the roadway. The intent is to eliminate defined boundaries for both the newly created parks and the development site. This produces the double benefits of a de-facto larger park, and the perception of a larger site from which the buildings rise.

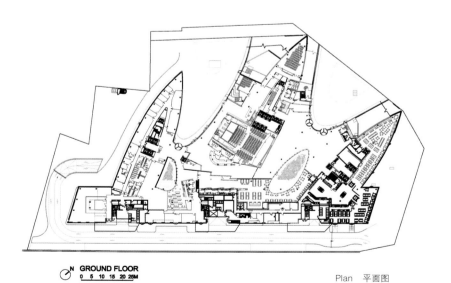

N GROUND FLOOR
0 5 10 15 20 25M

Plan 平面图

项目简介

这是微软集团最新的欧洲总部设计项目。Bouygues Immobilier 和 Generali 集团共同开发的新楼址位于巴黎市郊 Issy les Moulineaux 的塞纳河畔,占地 46 000 平方米,并享有塞纳河的壮丽景色。这一项目由三座大楼组成,每一座占地 15 000 平方米,它们由庭院、零售店和一个三层的停车场连接。

设计意图

这一发展区的设计是由三个体量的类别决定的。

长条体块:沿着 SCNF 软磁轨道放置一个矩形棱镜,设定了一个城市边界,并定义了这一发展区。这个长条体块与周围建筑形成正交结构,它还为 SCNF 的活动提供了视觉缓冲区,透过它的透明玻璃在视觉上与 Issy 区形成视觉联系。

顶楼:三座精心琢过的体块矗立在长条体块前,它们延伸到塞纳河边并在整体布局中独立呈现。它们中的两个横断外围,另一个则自由发展。这些形态可以唤起回忆并使它的特征明朗化。它们充满活力的造型寓意着动态感,这与外围的静态感截然相反。浅弧形的玻璃表面从长条处开始被分离,并又延伸至码头处时交汇于一点。由此产生的空间开阔了河边的视野,使人们在公路和码头都可以看到它们。

底层:裙房延伸部分是一个有机和犹如起伏波浪的结构,大楼在其基础上拔地而起。这种新的地势或人造地形隐藏了大楼底部的大部分功能,同时实现了意想不到的布局设置。裙房和码头沿线地面、附近的公园以及在办公楼和铁路之间新规划的公园融为一体,将所有的物业协调在一起。这样设计的目的是为了消除新建公园和新社区的界限。事实上它产生了双重收益——既创造了更大面积的公园,又为矗立在公园上方的建筑营造了绝妙的视觉效果。

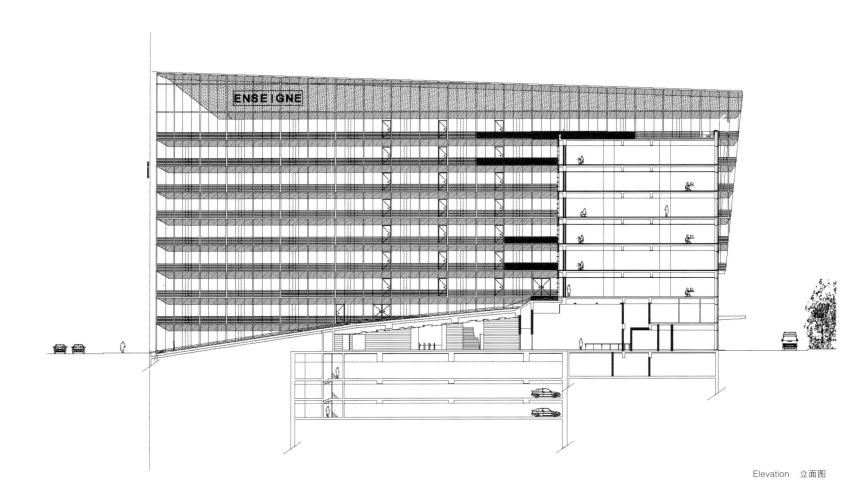

Elevation 立面图

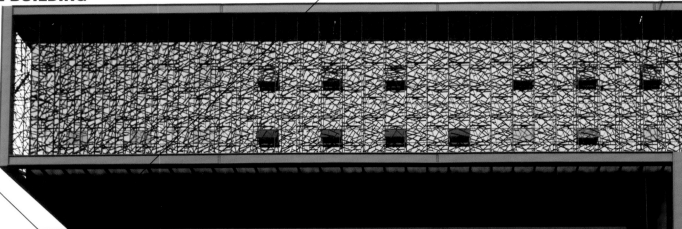

SOUTH NINGBO BUSINESS DISTRICT
宁波南部商务区

Architects: MADA s.p.a.m.
Business planning: SPAMALL Investment Planning Ltd.
Collaborators: Architecture Design Institute of Yinzhou, Ningbo,
Zhongding Architecture Design Institute, Ningbo
Location: Ningbo, China
Site area: 84,481 m^2
Gross floor area: 366,564 m^2

设计机构：马达思班建筑设计事务所
商业策划：思班奥投资策划有限公司
合作单位：宁波鄞州建筑设计院、
　　　　　宁波中鼎建筑设计研究院
项目地点：中国宁波市
总用地面积：84 481 平方米
总建筑面积：366 564 平方米

General Layout Design
In the overall design, Water Street will be the center of Yinzhou New Town,
positioned as a business office area. It will be the business activity and exchange
center of Yinzhou in the future. The building functions as headquarters, coupled
with other supporting functions such as cultural exchange, business services,
apartments, and hotels.
The planning programs put emphasis to the matching of relevant architectural
features to share diversified building types and spatial layout form and a 7 ×
24(7 days a week, 24 hours a day) wonderful vibrant area.
In the eastern half of the plot, there is a public service corridor which holds
business activities, sharing information center as well as the public platform
together. Architects set the commanding heights of buildings in this area, the
building will enrich iconic architectural image.
In the western half of the plot, the riverside activity spaces are organized around
the river mainline. It is going to creating a commercial and cultural pedestrian
street. Besides, combined with the indoor and outdoor space of the architectural
monomer, the diversified water landscapes are dotted here.
This program emphasizes on the creation of a humanizing public space,
designing a pedestrian landscape entertaining zone throughout No.1 to No.4 plot
in the north. And around this mainline, there is gradually forming a commercial
pedestrian street, a waterfront landscape gallery and a sightseeing channel over
the river.

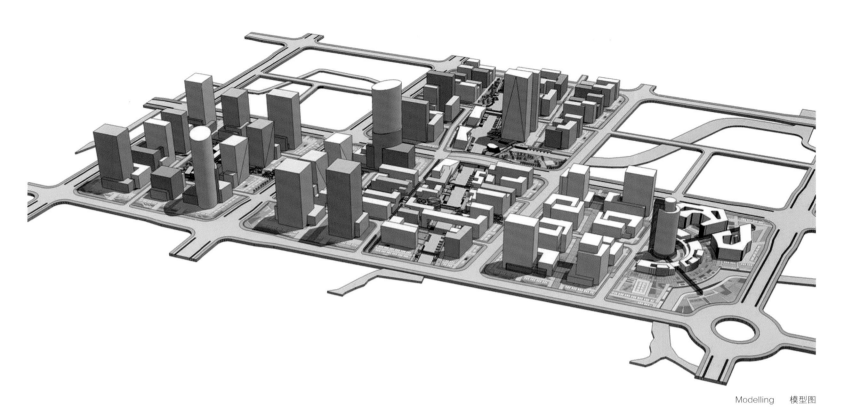

Modelling　模型图

Architectural Design

The southern business district in Yinzhou New Town is one of the ten major functional blocks to carry out the strategy of "Central Improvement".

The southern business district will be the business activity and exchange center of Yinzhou in the future. The design is typically for the core area of business district—Water Street, which enjoys extremely important role and value.

The core area of the eastern new town is located in the junction region between the new town and the old town. According to the planning, the core area will be the key to the future development of Ningbo City which shoulders the responsibility as new town with the old town. The core area will focus on the development of business office and supporting functions, coupled with the characteristics of neighborhoods and communities in the water town, becoming the logo and image for the future development of Ningbo City.

総体布局设计

在整体设计中，水街将成为未来鄞州新城区的中心区域，用地功能定位为商务办公区。它是鄞州未来的商务活动和交流中心。建筑功能以办公为主，同时兼容文化交流、商业服务、公寓、酒店等配套功能。

方案重视建筑特征的匹配，形成多元化建筑类型和空间布局形态以及"7×24"（每周7日，24小时全天候）模式的精彩活力区。

在地块的东半部，一条公共服务走廊将商业活动、共享信息中心以及公共交流平台等整合起来。设计师将建筑台的制高点设在该区域，形成富有标志性的建筑形象。

在地块的西半部，以河流为主脉，组织滨河活动空间，打造出一条商业文化步行街。此外，结合建筑单体的室内外空间，点缀多样化水体景观。

方案重视人性化公共活动空间的营造，设计一条贯穿北部1至4号地块的步行景观休闲带。规划河流贯穿全区，并以此为主线逐渐形成滨河商业步道、滨水景观画廊和观光通道。

建筑设计

地处鄞州新城的南部商务区是实施城市"中心提升"战略的十大重要功能区块之一。

南部商务区是鄞州区未来商务活动和交流的中心。而本次设计是商务区的核心区——水街部分，其作用和价值极其重要。

东部新城核心区位于新城和老城交界处。按照规划，核心区是宁波市未来发展的关键，将与老城一起承担新区的职能。核心区着力发展商务办公以及相关配套功能，再配以水乡特色的邻里社区，将形成宁波市未来发展的标志和形象。

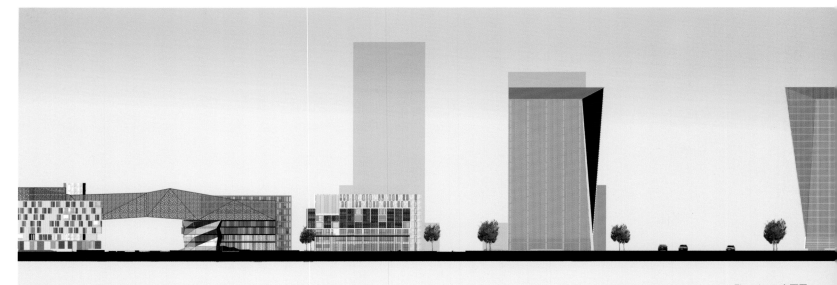

Elevation 立面图

JINSHANGU CREATIVE OFFICE PARK

金山谷创意产业园

Architects: Guangzhou Hanhua Architect+engineers CO.LTD.
Location: Guangzhou, China
Site area: 130,000 m²

设计机构：广州瀚华建筑设计有限公司
项目地点：中国广州市
总用地面积：130 000 平方米

Jinshangu Creative Office Park blends many elements together, including IT-based modern offices, shops, entertainment facilities, as well as the first-class residence and open spaces. In the 24-hour operation complex, people live, work, play with abandon, seek the comfort of urban life in the high-quality environment. Here, people can live in luxury private residence, shopping and playing along the linear park, communicating with visitors. The outsiders can also come to the green ecological environment via various convenient traffic methods to work and enjoy entertainment. Jinshangu, as a complex community, provides a complete way of life.

金山谷创意产业园由诸多功能元素组成：以 IT 为基础的现代化办公空间、商铺、娱乐设施以及品质一流的住宅和开放空间。在这座 24 小时无休的综合体中，人们可以尽情生活、工作、游玩，在高品质的环境中寻找都市生活的舒适感。在这里，人们可以居住在豪华的私人居所，购物，沿着线形运动场游玩以及与游客交谈。外来者也可以乘坐多种便捷的交通工具抵达这个绿色生态环境，在这里工作、娱乐。金山谷作为一个综合的社区，提供了完整的生活方式。

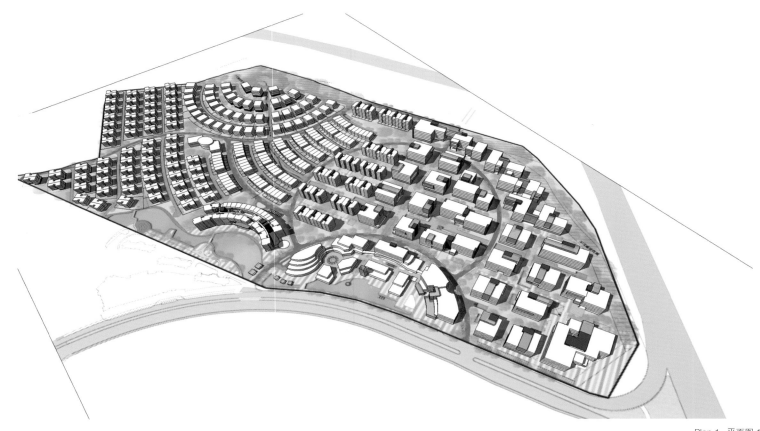

Plan 1 平面图 1

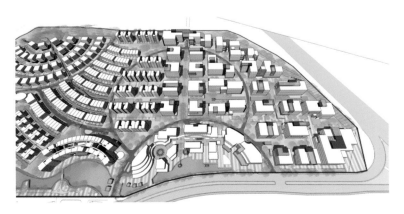

Plan 2 平面图 2

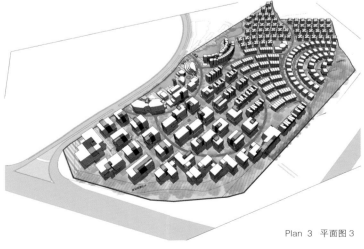

Plan 3 平面图 3

CHEUNG KONG CENTER
长江集团中心

Architects: Leo A Daly company
Location: Hong Kong, China
Site area: 9,600 m²

设计机构：Leo A Daly
项目地点：中国香港特别行政区
总用地面积：9 600 平方米

Project Introduction
The Cheung Kong Center is 283 m-tall and covers about 9,631 m². It is situated at the intersection of Garden Road and Queen's Road.
The design organizes the steep topography of the site on three levels, each with its own distinct character.

The Lobby
The base is designed as a fully transparent, two-level lobby to be entered from both the north and the south. Glazed with high-performance Viacron with 100 percent light transmittance, the large enclosing walls were custom-fabricated.

Structural System
Structural engineers designed an inner core of concrete and an outer structural frame of concrete-filled circular steel tubes placed on a 7.2 m grid. Flexible, column-free floor plates averagely reach 1, 720 m² in area.

Office Environment
To admit as much natural light as possible, floor-to-ceiling office heights of 3 m are as standard.

Curtain Wall as Tapestry
The design team created facades of interwoven grids and spatial layers for the skin of the tower which produced a tapestry like effect.
The exterior column bays are subdivided with vertical mullions spaced on 2.4 m span. Each bay, in turn, is horizontally organized into four window components.

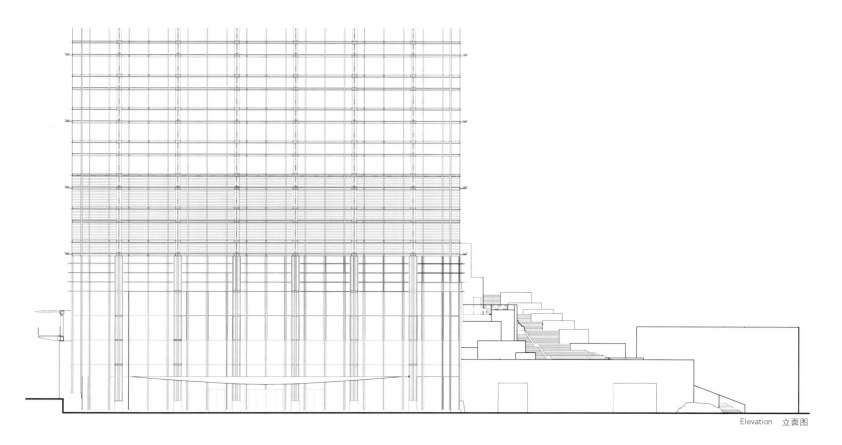

Elevation 立面图

Layers in Light

The exterior surfaces are all fabricated of stainless steel with a "linen" finish because of its greater reflectance.

To ensure its distinctive contribution to Hong Kong's illuminated shoreline at night, more than 12,000 emitter lenses placed at the intersections of all horizontal and vertical mullions of the facades appear as individual "stars" of light.

Urban Poise

At street level, the Cheung Kong Center seamlessly merges its dramatic lobby spaces with the exterior civic amenities created around it.

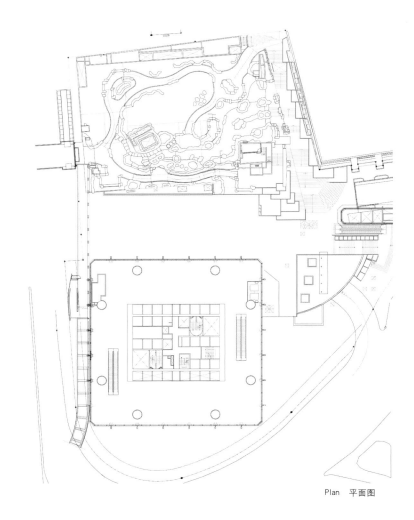

Plan　平面图

项目介绍

长江集团中心高 283 米,占地 9 631 平方米,位于香港花园道与皇后大道交界处。设计将地形陡坡巧妙地整合为三个层次,每层阶梯都具有显著的特色。

前厅

底部为全透明的双层大厅,游客可从南北两侧进入。围墙由完全透光的高性能釉面材料 Viacron 制成。

结构系统

结构工程师将混凝土内芯和填满混凝土的外层钢管构架置于 7.2 米的栅格上。灵活、无立柱的楼板平均面积达到 1 720 平方米。

办公环境

为了尽可能多地收到自然光线,大厦均以 3 米的天花板高度作为标准。

织锦幕墙

设计团队创造了一个相互编织的多层立面,形成了织锦般的视觉效果。外部立柱跨度被间隔 2.4 米的窗户框架分开,每个跨间间水平排列了四个窗户。

光线层面

为了实现较高的反射率,大厦外表面全部用不锈钢装饰为"亚麻"质地的视觉效果。为了确保其对香港夜间海岸线景观的照明,长江集团中心将 12 000 多个发射器镜头放置在立面所有水平和垂直竖框的交界处,形成了一片星光。

城市姿态

在街道层面,长江集团中心将其引人注目的大堂与外部公共设施无缝融合。

SHENZHEN LOGAN CENTURY PLAZA
深圳龙光世纪大厦

Architects: AECOM
Site area: 79,800 m²
Location: Shenzhen, guangdong, China

设计机构：AECOM
总用地面积：79 800 平方米
项目地点：中国广东省深圳市

This project is located in the northeast corner of Central Square in Shenzhen's Bao'an District, facing the Municipal and Sports Center. It enjoys views of the central park and ocean scenery towards the southwest. The general plan calls for two towers connected by a retail base. An ocean-view corridor is formed between the two towers and buildings along the north and south blocks. The two towers are designed with a harmonious discipline, and offer rich visual variations when viewed from different angles. The site is divided into six parts. Tower buildings are placed at the northwest and southeast corners to define the urban center, to reduce visual interference between them, and to open up views to the city center. The other two corners on the site are made into plazas: on the northeast corner is a high-traffic commercial square and on the southwest a quiet business square with a view to a central landscaped area.

　　该项目位于深圳市宝安区中心广场的东北角，与行政中心、体育中心遥遥相对。它将西南方位的中心公园和海景尽收眼底。总体规划将该大厦设计为两栋塔楼，由一座商业区底座连接。建筑师在两座塔楼与沿着南北街区的建筑群之间建造了一条全海景观景廊。两座塔楼严格遵守和谐准则，从不同角度营造出丰富的视觉盛宴。该大厦被划分为六部分，塔楼设置在西北角与东南角，从而明确了其城市中心的形象特征，减少了相互之间的视线干扰，营造出城市中心区的极致景观。位于两座塔楼对角处的是两个广场：东北角为熙熙攘攘的商业广场，西南角则是配有中心绿带景观且相对安静的商贸广场。

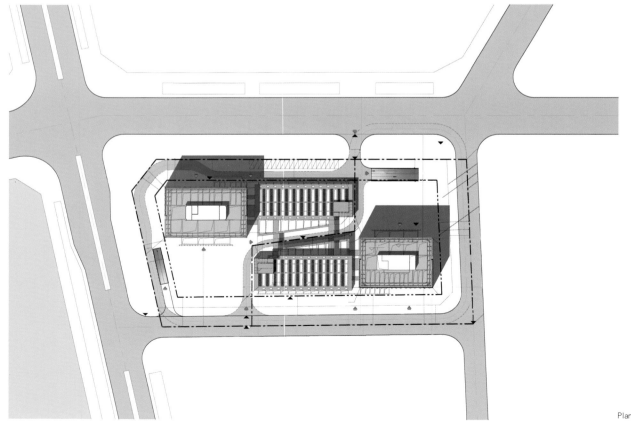

Plan 平面图

SHANGHAI HUANGPU CENTER
上海黄浦中心

Architects: Shanghai ZF Architectural Design Co., Ltd.
Location: Shanghai, China
Site area: 22,900 m²

设计机构：上海中房建筑设计有限公司
项目地点：中国上海市
总用地面积：22 900 平方米

Shanghai Huangpu Center is the location of the Shanghai ZF Architectural Design Co., Ltd. as well as a success practice of "the overall design" and "full-range control". The project is located at the old Simon area in Huangpu District, facing Zhonghua Road green belt to the east, and the New World to the west. The site area about 5,700 m², the floor area above ground is about 22,900 m², FAR is 4.0, 20 levels on the ground, 2 levels underground, standard floor area is about 12,000 m², standard level is 4 meters tall, the completion date was in June, 2008.

Paying Attention to the Surrounding Landscape Value, Harmonizing the Relationship between City Spaces

The surrounding of the base is complex, and the plan reduces the building density and increases the green coverage rate as far as possible. The green setting echoes with green belt of the old city. Architectural space, shape relations as well as the streamlined organization are taken into consideration in the office building phase 2. The building with steady color is simple and regular and makes full use of the landscape value of the New World on the east and Huangpu River on the west to reduce the sight obstruction from the existing high-rise residential building on the south, and becomes the striking scenery in the area.

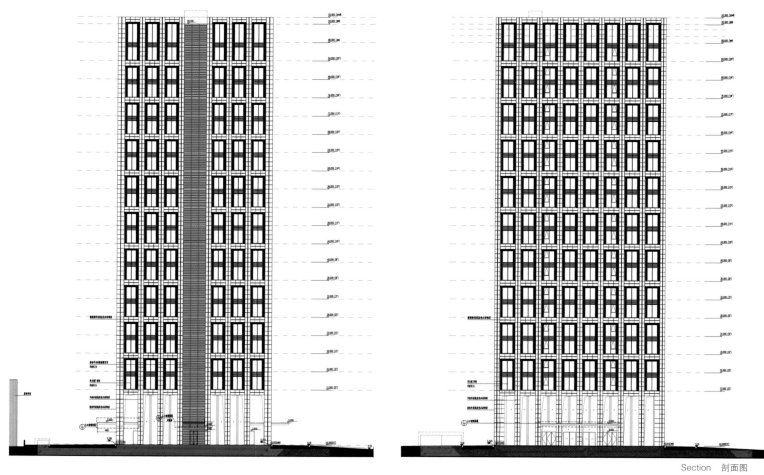

Section 剖面图

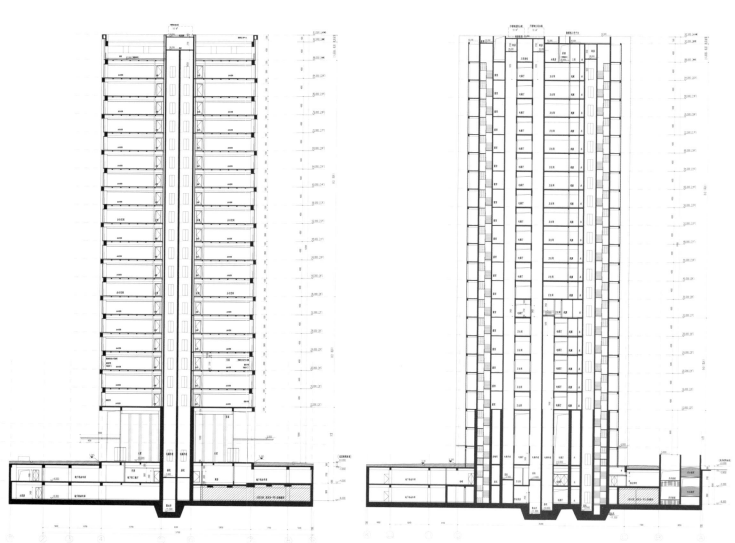

Elevation 立面图

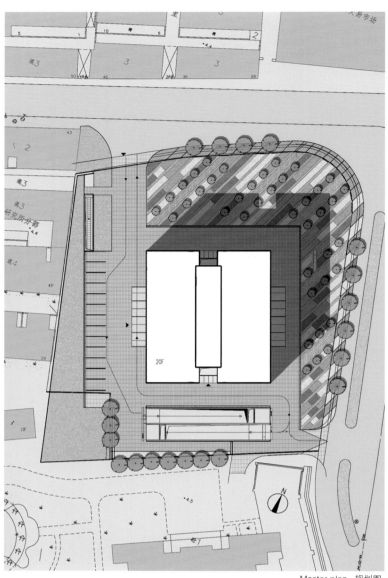

Master plan 规划图

Simple and Modern Architectural Style, High-efficient and Flexible Function Feature

Building adopts the unit-type curtain wall made from glass and stone which is concise, modern and delicate. The ground lobby with floor-to-ceiling transparent glass, the revolving door and glazed glass canopy forms a transparent and scale-appropriate entrance space. The outward opening windows with the fixed shading louvers are installed on the both side of the glass curtain wall. The fixed louvers are installed on the device platform outside whose north and south ends are concave to highlight the gradation of the building.

Harmonious Interior Design Strengthening Modern Office Quality

Interior design focuses on the coordination with the style of the building facade and the materials. The entrance lobby wall, floor and roof are made of stone or artificial stone to blend the interior and exterior of the building into one harmonious whole. Elevator lobby wall and top finishes use brushed stainless steel to make a sharp contrast to the stones. Office spaces adopt a standard module form. The ceiling lighting, smoke sensor and spray head are unified into an integrated lamp belt to highlight the concise design concept and the consistence with architectural style.

Utilizing the Texture Rhythm Efficiently, Using Environment to Contrast the Modern Atmosphere

The outdoor landscape visual center is the green ramp along the road. The design of the green ramp pays attention to the coordination between the plane texture and the facade. The strip-shaped gardens arranged densely incline from inside to outside with a strong rhythm and a sense of direction, and echo with the vertical lines of the construction. The building breaks out of the ground, and the night landscape lighting reinforces this effect.

　　上海黄浦中心是上海中房建筑设计有限公司的所在地，也是遵循"整体设计"和"全程控制"战略的一次成功实践。基地位于黄浦区老西门地区，东邻中华路环城绿带，西与新天地咫尺之遥。用地约 5 700 平方米，地上建筑面积约 22 900 平方米，容积率 4.0，地上 20 层，地下 2 层，标准层面积约 12 000 平方米，标准层高 4 米，竣工日期为 2008 年 6 月。

　　注重周边景观的价值，协调城市空间的关系
　　基地周边环境复杂，规划尽可能降低建筑密度，提高绿化覆盖率，绿化设置与老城区的环城绿带相呼应。建筑空间、形体关系以及流线形组织均与二期办公楼统一考虑。建筑单体色彩稳重、简洁方正，并充分利用东西两侧新天地和黄浦江的景观价值，减少南侧已建高层住宅所造成的视线干扰，成为该区域引人注目的风景。

　　简洁现代的建筑风格，高效灵动的功能特性
　　建筑采用了玻璃与石材相结合的单元式幕墙，简洁现代又不失精致。底层大堂采用通高透明的玻璃，结合旋转门和彩釉玻璃天棚，形成通透、尺度适宜的入口空间。在玻璃幕墙两侧选用了外开式窗户，并设固定遮阳百叶；南北两端凹进的设备平台外侧也安装了固定百叶，突显建筑的层次。

　　风格协调的室内设计，强化现代办公的品质
　　室内设计注重与建筑立面风格和材料的协调。入口大堂的墙、地板、屋顶均采用石材或人造石材，使建筑内外浑然一体。电梯门厅的墙、顶饰面采用拉丝不锈钢，与石材形成强烈对比。办公部分采用了标准模块的形式。吊顶照明、烟雾感应器、喷淋头等设备统一设计为集成式灯带，突出了办公空间简洁的设计理念，并与建筑风格相一致。

　　肌理节奏的有效运用，以环境烘托现代氛围
　　室外景观的视觉中心为沿路绿坡。绿坡的设计讲究平面肌理与建筑立面的协调。密集排列的带状花池由内向外倾斜，具有强烈的节奏和方向感，并与建筑的竖向线条相呼应，形成了建筑破地而出的向上态势，夜间的景观照明更加强化了这一效果。

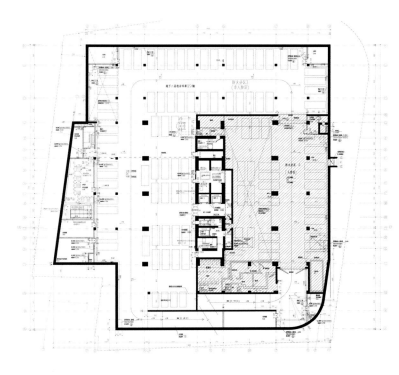
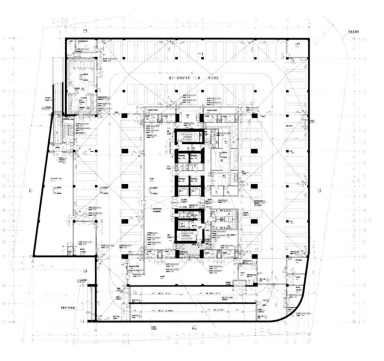

Ground Floor Plan　首层平面图

XI' AN TELEVISION BROADCASTING CENTER
西安广播电视中心

Architects: MADA s.p.a.m
Associate architects: The Institute of Architectural Design and Research, Shenzhen University; Radio Film &Television Design and Research Institute
Location: Xi'an, China
Site area: 81,117 m²

设计机构：马达思班建筑设计事务所
合作设计单位：深圳大学建筑设计研究院、
　　　　　　　中广电广播电影电视设计研究院
项目地点：中国西安市
总用地面积：81 117 平方米

The project has inherited the generous and grand features of the culture of the Han and Tang dynasties in Xi'an. The buildings highlight and exaggerate the dimension concept under the given building area condition. The various functions in the overall architecture are closed by a symbolic wall whose boundary is entity-oriented. The imaginary wall expresses the historical sense of the wall in Xi'an, but it emphasizes the modernization of the materials and the construction form. The facade is divided into an inner layer and an outer layer: The inner layer embodies the inner functions of the buildings while the outer facade is unified in the transverse lines of uneven density. The clay louvers with different gap widths are set on the windows, which reflect the weight sense of classical buildings and lightsome feature of modern architectures. Some special functions in the sculptural art wall are designed to be out of the wall such as the "Media Theater" at west which jumps from the wall to form the core icon of the public space on the west side and echo with the city. The roof of the atrium introduces the grant dimension of Chinese classical hall and the fold logic of wood structure and chooses the rational geometrical triangle. These triangular truss planes, some lower, some higher, some flat, some inclined, not only reduce the need for columns, but also maximize the artistic charm of the interface between the roof and the ground. By light filtering and shading, the approach has completed the creativity of "sunlight media hall".

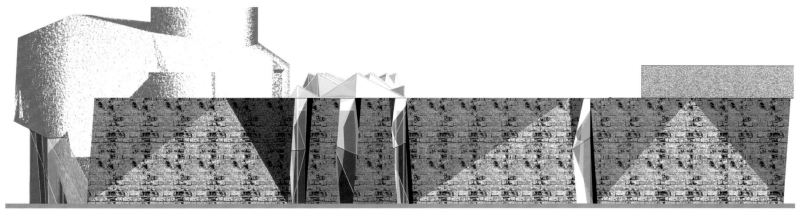

Section　剖面图

Elevation　立面图

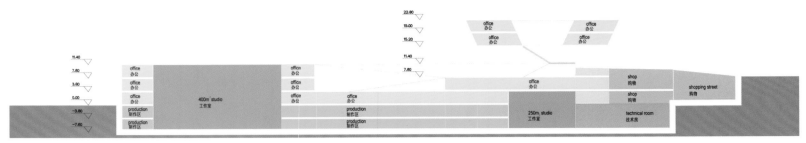

Section 1 剖面图 1

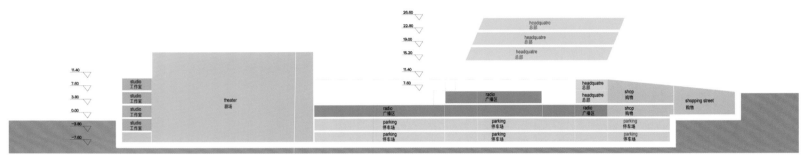

Section 2 剖面图 2

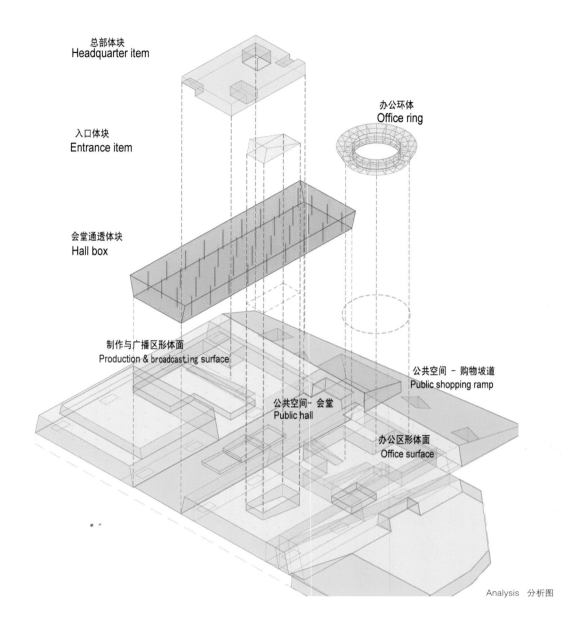

总部体块
Headquarter item

办公环体
Office ring

入口体块
Entrance item

会堂通透体块
Hall box

制作与广播区形体面
Production & broadcasting surface

公共空间 - 购物坡道
Public shopping ramp

公共空间- 会堂
Public hall

办公区形体面
Office surface

Analysis 分析图

本项目承袭了西安汉唐文化中惯有的舒展大度、简洁雄浑的特质。建筑群突出并夸张了现存建筑面积条件下的尺度概念。整体建筑的诸多功能由一道具有象征意义的"墙"围合起来，墙体的边界以实体感为主。这座虚构的城墙虽透着西安城墙的历史气质，却强调了用材及构造方式的现代化。建筑立面分内外两层：内皮为建筑物内部功能的真实体现，外立面则被统一在疏密不同的横向线条中。开窗处设置有不同宽度间隙的陶土百叶窗，既体现了古典砌筑方式的重量感，又表达了现代建筑的轻盈属性。在这具有雕塑感的艺术墙体中，有一些特殊功能被设计成"出墙"状态，譬如西侧的"媒体剧场"，即是从墙中凸显出来，形成西侧公共空间的核心标志物，并与城市空间遥相呼应。建筑中庭的屋盖在概念上引用中国古典殿堂的宏大尺度及木结构的折面逻辑，选择了合理的几何三角形。这些三角形构架面有高有低、有平有斜，不但减少了空间中所需的柱子数量，又最大化地体现了屋顶与地面形成的空间界面的艺术震撼力，这一手法在经过科学遮光、滤光处理后，便完成了"阳光媒体殿堂"这一创意。

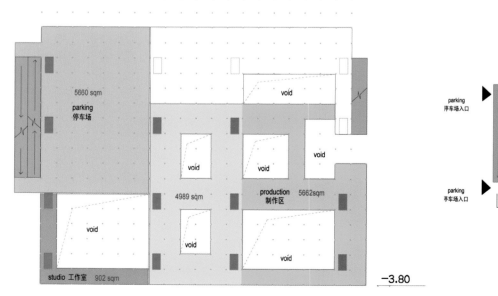

Underground 1st Plan 地下 1 层平面图

1st Floor Plan   1 层平面图

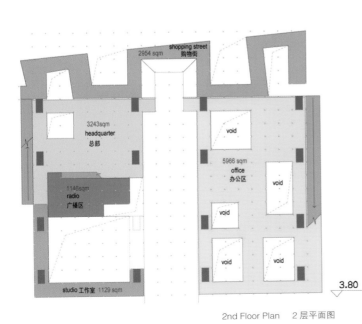

2nd Floor Plan　2层平面图

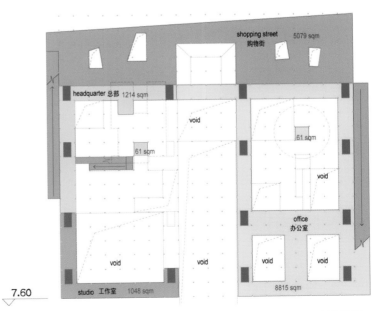

3rd Floor Plan　3层平面图

JINAN HANYU FINANCIAL CENTER
济南汉峪金融中心

Architects: DAO
Location: Jinan, China
Site area: 890,000 m²

设计机构：DAO 国际设计集团
项目地点：中国济南市
总用地面积：890 000 平方米

The project is the core center of Hanyu Financial Center, located on the south side of Jingshi Road, west side of Fenghuang Road, north side of high-tech IT headquarters office in Jinan. The architects adopted the concept of "WEAVE" to complete the project design:

• "W" means water. The architect divided the site into three blocks according to their characteristics and named them as water origin, water tide and water life accordingly.
• "E" means efficiency. The design emphasizes on the functionality, economy and feasibility, creates convenient underground space, transport organization and takes advantage of the high efficiency of the super high-rise.
• "A" means accessibility. The design encourages interactive and open places, provides reasonable scale for human activities, and promotes walking instead of driving.
• "V" means vibrancy. The architect attempts to form a vivid city with tension and energy.
•The very last "E" means ecology. The architect focuses on the relationship between the architecture and the surrounding environment and emerges the modern city life into the nature.

　　本案位于济南市经十路南侧，凤凰路以西，北与高新区 IT 总部基地相望，是汉峪金融中心的核心区。建筑师以"WEAVE"理念来设计此项目。
　　建筑师根据地块的特点，将项目划分为三个区域，并分别命名为泉源、泉潮、泉生—— W 即为水。
　　设计强调功能性、经济性和可行性，创造出便捷的地下空间和交通组织，体现出超高层建筑的高效性—— E 即为高效。
　　设计提倡开放互动的空间，合适的人动尺度，鼓励步行—— A 即为可行性。
　　建筑师试图打造一个充满活力和张力的城市空间—— V 即为活力。
　　建筑师强调建筑与周边环境的关系，将城市生活融入自然—— 最后的 E 即为生态学。

East Modelling 东立面模型图

South Modelling　南立面模型图

JIANGXI NANCHANG INDUSTRIAL PARK OF YONYOU GROUP
江西用友集团南昌产业园

Architects: The Z&Z Studio in The Institute of Architecture Design & Research, Shenzhen University
Project team: Zhong Zhong, Zhong Botao, Yuan Lei, Feng Ming, Xiong Qiang, Cen Min, Gu Zhihui
Location: Nanchang, China
Site area: 409.700 m²
Gross floor area: 480,757 m²
Building height: 8 levels, a maximum height of 38.45 m

设计机构：深圳大学建筑设计研究院 Z&Z 工作室
设计团队：钟中、钟波涛、袁磊、冯鸣、熊强、岑敏、辜智慧
项目地点：中国南昌市
总用地面积：409 700 平方米
总建筑面积：480 757 平方米
高度：8 层，最大高度 38.45 米

Project Background
Guided by the principle of ecological and natural, practical and concise, modern and national, economic and durable, the project was designed to be a garden-style harmonious software industry park in coordination with natural environment.

Site Analysis
The industrial park faces Wangcheng Avenue on the northeast, Jiangxi Vocational and Technical College of Industry and Trade on the opposite side, Nanchang University City on the southeast across the Changzhang Expressway. There are urban roads around the site.
The site area is about 40 hectares and the relative altitude is less than 20 meters.

Design Strategy
1. Earth Landscape
The construction breaks up the whole into parts, fuses into the environment, and hides in the mountains to achieve the goal of "green Nanchang".
2. Ecological Garden
Combining with natural relative elevation and the local "landscape + soil sealing" pattern, the design fully protects the landform and vegetation to create a landscape park.
3. Low-carbon Design
The design emphasizes on "low-impact development" and "passive energy-saving", combines the entire ecological layout with the green design of building units to realize the low-carbon ecology in whole building life cycle.
4. Two-star Green Building
The design adopts many technologies such as air conditioning cold and heat sources, water source heat pump, LED energy-saving light source, light-guide lighting technology to come up to the national standard of "two-star green building demonstration zone".

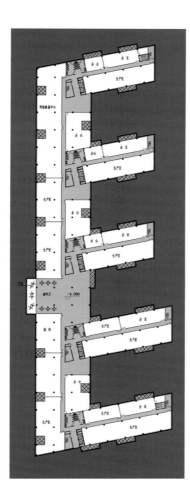

Underground 1st Plan 地下 1 层平面图

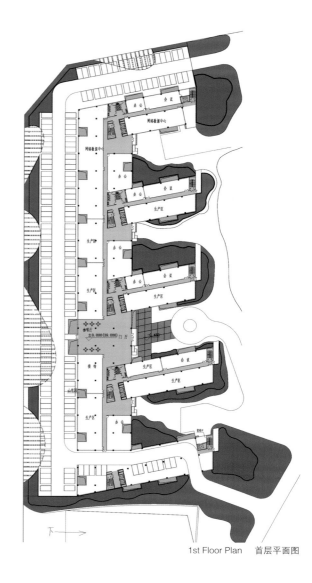

1st Floor Plan 首层平面图

Building Design
1. The Soil-sealing Landscape
This design uses the "landscape" and "soil sealing" pattern.
2. Mixed Development
Designers created "hybrid workspace" to realize circulatory dynamic utilization of space resource.
3. Common Space
A large amount of common spaces such as sky gardens, roof gardens and inner courtyards create cultural communication center for the building.
4. Jiangxi Gene
The design respects for local culture and tries to fuse "Jiangxi gene" into the project.
5. Digital Construction
The double-skin structure of the building shows the pixel pattern, colorful facade texture and strong identifiability, and contacts the skin with ecological energy-saving technology to achieve good ventilation and sunlight reflecting.

Section 1 剖面图 1

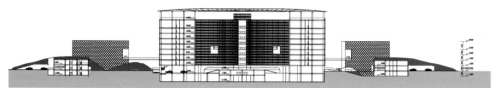

Section 2 剖面图 2

项目背景
本项目在贯彻"生态自然、实用简洁、现代民族、经济耐用"方针的指导思想下，建设与自然环境协调一致的花园式软件产业园。

用地分析
产业园东北毗邻望城大道，对面是江西工业贸易职业技术学院，东南隔昌樟高速可远望南昌大学城，周边均设有市政道路。
基地面积约 40 公顷，相对高度约 20 米。

设计策略
1. 大地景观
建筑化整为零，融入环境，隐入山林，实现"绿色南昌"的目标。
2. 生态田园
结合地形的自然高差，采用局部"地景 + 覆土"模式，充分保护地貌、植被，营造一个天然公园。
3. 低碳设计
设计强调"低影响开发"和"被动节能"，将"整体生态布局"与建筑单体的"绿色设计"结合，实现建筑全生命周期的低碳生态。
4. 绿色二星建筑
综合采用空调冷热源、水源热泵、LED 节能光源、光导照明等技术，以期建成后达到国家"二星级绿色建筑示范区"标准。

建筑设计
1. 地景覆土
本设计采取"地景 + 覆土"模式。
2. 混合开发
设计师塑造出"混合型工作空间"，以实现空间资源的循环动态利用。
3. 共享空间
大量的空中花园、屋顶花园、内部庭院等共享空间为建筑营造了一个文化交流中心。
4. 江西基因
设计尊重地方文化脉络，力图把"江西基因"融入项目。
5. 数字化加工
建筑采用双层表皮构造，展现像素化图形、多彩的立面肌理、强烈的可识别性，同时将表皮与生态节能技术相联系，以达到良好的通风和反光条件。

TWO E-COM CENTER
两个电子商务中心

Architects: Arquitectonica
Landscape architect: DQA Environmental Design
Interior designer: M Concept Design
Location: Manila, Philippines
Total area: 73,500 m²
Photographer: Pathfinder

设计机构：Arquitectonica
景观设计：DQA 环境设计
室内设计：M 概念设计
项目地点：菲律宾马尼拉市
总用地面积：73 500 平方米
摄影：Pathfinder

The Two E-com project is a 73,500 m² office development located in the mall of Asia Complex, Pasay, Metro Manila. It has 3 podium parking levels and 12 office levels, with some ancillary retail at the ground and podium level.

The project is part of a larger 67-hectare master plan located around the 386,000 m² Mall of Asia. Other components in the master plan include the 46,650 m² SMX Convention and Exhibition Center; the 15,000-seat Mall of Asia Arena; the 74,300 m² One E-com center; and future hotels and serviced apartments.

The project has been designed to cater for the growing BPO/call-center industry in Philippines, which requires large, continuous office floor plates. Realizing that the podium parking would also virtually occupy the whole site, designers wanted a green podium deck which would be easily accessible to the public.

The podium will contain multiple entry lobbies at ground level, and will be lined with single & double height street-retails around its perimeter arcade. Tenancies will include convenience stores, F&B outlets, banks & ancillary services to support the large office population.

The composition of the Two E-com project is a study in contrasts: between its white, gridded, orthogonal perimeter, and its faceted, dark inner facade; between its sculptural mass and the cuts and openings in its form. The focus on the archway space balances the impact of the building massing, transforming its bulk into a dynamic composition, and giving the client the architectural identity they desired for their headquarters.

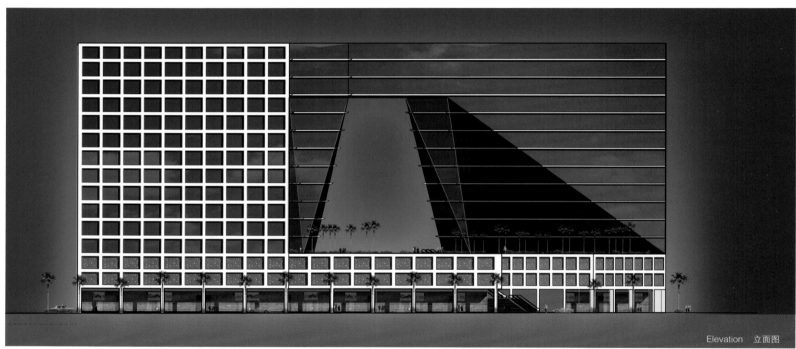

Elevation 立面图

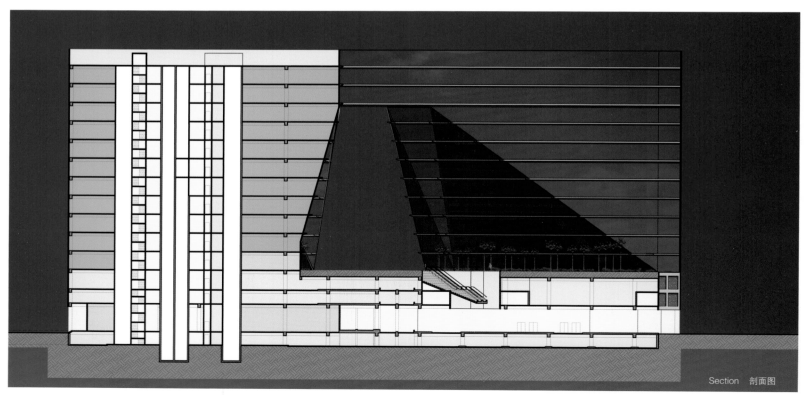

Section 剖面图

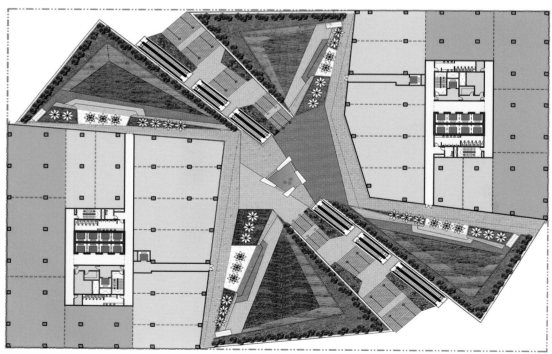

LEGEND

■ OFFICE

■ RETAIL

□ PUBLIC CIRCULATION

■ CORE / MEP

■ PARKING

Plan 1　平面图1

Plan 2 平面图 2

两座电子商务大厦位于马尼拉南部城市帕赛市的亚洲商业购物中心，为占地 73 500 平方米的办公空间。其中包括 3 层停车场地、12 层办公空间及一些附设的地面层和高层零售设施。

该项目是 67 公顷土地总体规划的一部分，位于 386 000 平方米的亚洲购物中心周围。总体规划中的其他部分包括 46 650 平方米的 SMX 会展中心、有 15 000 个席位的亚洲购物中心活动舞台、一个 74 300 平方米的电子商务中心以及未来酒店和服务型公寓。

该项目的设计宗旨是为了适应菲律宾日益增长的商务流程外包业务、传呼中心服务行业，这些均需大型的、连续的办公场所。设计师计划建造一个绿色环保的停车场所，停车场地将占据整个楼面，方便进入公共区域。

大厦将包括多个地面入口大厅，在街道的拐弯处设有单层和双层的零售店铺。出租地块包括便利店、F & B 零售商、银行与辅助服务设施，以维持大量的办公人员的生活。

这个两座电子商务大厦方案有着鲜明的对比：白色网格状的直角形周界与多面的黑色内立面；大型的雕刻建筑体块与形式上适当的切口。拱门空间有效地平衡了整个建筑体块的影响，将其转换成一个动态的组合，形成客户所需要的建筑形象。

TEMPO SCAN TOWER
TEMPO SCAN TOWER

Architects: Arquitectonica
Location: Kuningan, South Jakarta, Indonesia
Site area: 44,200 m²
Photographer: Eric Niemy, Jonathan Rukma

设计机构：Arquitectonica
项目地点：印度尼西亚雅加达市
总用地面积：44 200 平方米
摄影：Eric Niemy, Jonathan Rukma

Design Intent

Tempo Scan Tower rises from a prime site on Jakarta's Golden Triangle, in the heart of the leafy embassy district. The sculptural tower is iconic in form, stamping its presence as a Jakarta's landmark.

The tower is shaped by a skin that curves at the corners, creating a multidimensional presence. Each side is gently bowed, narrowing at the base, and broadening at the center, finally narrowing again at the roof.

The facade itself is a simple, high-performance IGU window wall system spanning floor to floor between slabs, both cost-effective and constructible. The exterior is articulated by vertical fins that stagger from one floor to the next. The entrance canopy is a streamlined shape clipped onto the building's corners, projecting dramatically over the port nearby.

The 30-storey building is pure in form and continuous in massing without interruption. The program contains almost 44,160 m² of gross floor area, predominantly Grade-A leased office space. The lowest 2 levels contain a spacious, interconnected program which combines a specialty restaurant and high-end "urban kitchen" food court on levels 1 & 2 respectively. Square size is planned in roughly 1,500 m². Core to wall dimensions are generously 10.5 m to 12 m clear spans. Office floors are split into two efficient zones served by 10 lifts, securing from access to the lobby from the basement levels.

The building is not served by a podium which keeps site clutter at a minimum, and allows the building clear orientation for access and enjoyment. Site coverage (KDB) is a generously 25%. Surrounding the building is a raised terrace that extends from ground-level restaurants, and permits a fashionable deck for outdoor café seating. Low-maintenance landscaping is maximized, including on-grade visitor parking that uses landscaped pavers to soften the overall effect.

Below the development is a four-level parking that provides Grade-A parking allowances for both Office and all F&B/Function occupants. For basement users, convenient and direct access to the lobby and lower Food court/Function floors is provided with 3 car park shuttle lifts.

设计意图

Tempo Scan 大厦位于雅加达市的黄金三角地段——绿树成荫的使馆区的核心区。雕刻状的外形是大厦的标志，这使它成为雅加达市的地标性建筑。

建筑由拐角处弯曲的表皮塑造而成，使其呈现多维度的样式。每一面均略微弯曲，底部狭窄，中间宽敞，最后在屋顶变窄。

立面是简约、高性能的 IGU 窗户墙系统，跨越各个楼层，成本合理且具有建设性。外表面由层与层之间的垂直鳍状物连接。入口的流线形天棚夹于建筑的角落，在港口附近非常引人注目。

30 层的建筑外形简约且连续。这一项目的总建筑面积约 44 160 平方米，主要用作甲级写字楼。最底下的两层是一个宽敞、连通的项目——将 1 层和 2 层相互独立的特色餐厅和高端城市美食广场融为一体。广场计划占地约 1500 平方米。墙的净跨度大约 10.5 米到 12 米。办公楼层被划分为两个区，安置了 10 部电梯，可确保从地下室安全进入大厅。

这个建筑未设置裙楼，这使得视野开阔清晰，同时明确了建筑定位，方便访问和娱乐。建筑密度（KDB）大约 25%。建筑周围是一个凸起的阳台，从一层的餐厅向外延伸，同时为户外咖啡座位区提供了一个时尚的服务台。需低成本维护的景观得到了最大化开发，访客停车场使用了园林铺料材料，使整体效果更加柔和。

建筑物地下是一个 4 层的停车场，为甲级写字楼的员工和餐厅居住者提供车位。对于地下室的使用者来说，建筑提供了 3 部停车场穿梭电梯，可以使他们便捷迅速地通往大堂和美食广场及功能楼层。

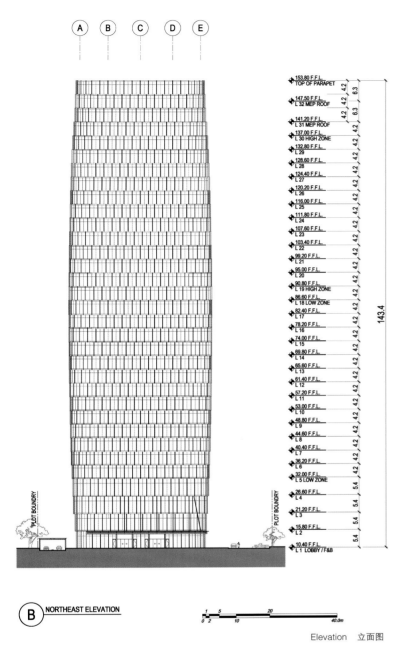

NORTHEAST ELEVATION

Elevation　立面图

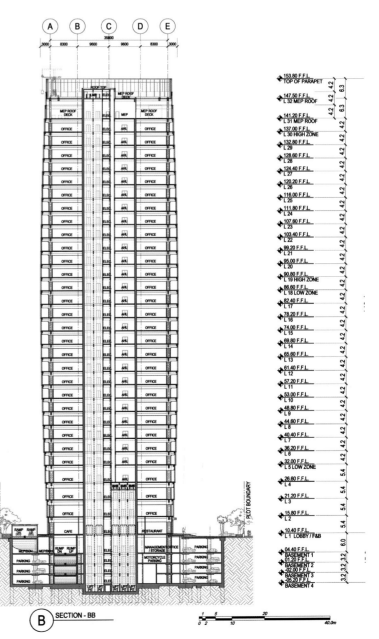

SECTION - BB

Section　剖面图

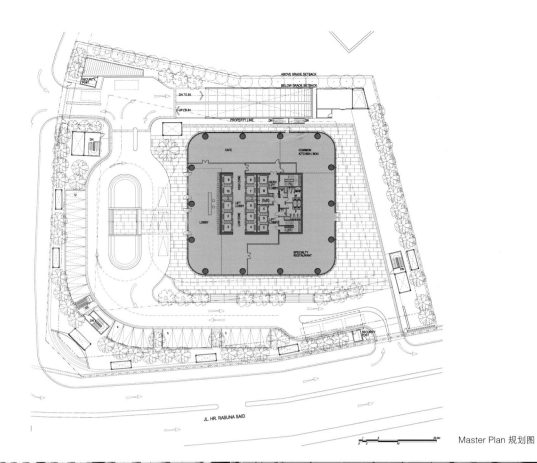

Master Plan 规划图

HONGQIAO AIRPORT TECHNOLOGY
虹桥临空科技产业楼

Architects: RIA International Research Institute of Architecture
Location: Shanghai, China
Site area: 70,149 m²

设计机构：RIA 国际都市建筑设计研究所
项目地点：中国上海市
总用地面积：70 149 平方米

Design Goals
The goal is to create the future, intelligent, and eco-harmonious environment office space.

Design Concept
Eco-harmonious Building
The project aims to achieve the perfect fusion of the architectural space and environment by using the most common building materials and technologies. It uses the traditional construction methods such as light-shading and light-resisting to reduce the room temperature changes influenced by climate of external environment. The natural ventilation and lighting lead the project to be a low-energy-consumption ecological building.

High-efficient Open Spaces Group
Six separate buildings are used as office headquarters with efficient and orderly construction layout. North and south groups for rent and sale are placed around two public open spaces. The groups with different uses contact each other through a central greenbelt, while separate from each other.

Intelligent, Garden-style Office Environment
The design introduces various intelligent systems such as disaster prevention system, management and communication system to ensure 24-hour humanized office environment. The green space combined with intelligent office environment provides the users with convenience, safety and efficiency from technology, and provides them with beautiful natural view.

Design Approach
Introducing Ecological Technique
The transparent building and natural lighting introduce the external green environment into the interior, saving artificial lighting energy cost. Roof greening is beneficial to the thermal insulation of the building, and vertical greening will effectively prevent and shade direct sunlight and reduce air conditioning energy consumption. The partial bottom overhead space, green atrium and openable skylights produce draft effect to achieve the natural ventilation and lighting. The solar energy collector which can adjust its angle following sunlight is set on the roof, making full use of natural resources.

The effective Space Layout
Buildings are orderly arranged and full of changes, strengthening the transparency of the landscape facing main road and the sense of sequence of the office spaces. Central green runs through the whole base, and combines the spaces on the ground and underground to produce spectacular spatial effect and interest. The south and north groups stretch around the two sunken public open spaces to provide more facilities and communication platform for the base. The pedestrian flows and vehicle flows separate to organize rational traffic line. The safe and comfortable central green walking system is arranged here.

Introducing Modern and Technological Green Environment
The modern and efficient communication systems such as IP telephone, satellite equipments etc. are imported. The floor height is a little tall to leave adequate space for the pre-embedded cable. The photovoltaic conversion system and a solar collector complete the building energy supplement. The waste water is used for flushing and irrigation, and the waste materials are used as fertilizer for park greening. Green rest atrium, balcony, and outdoor coherent ecological walking system, open space all form the comfortable office environment.

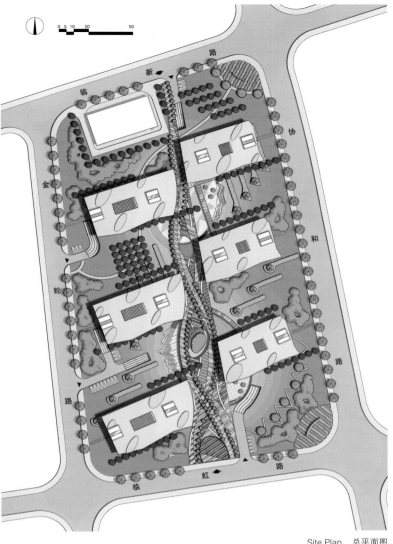

Site Plan 总平面图

设计目标
设计目标是创造未来型、智能化、与生态环境协调共生的办公空间。

设计理念
与生态相协调的建筑
建筑使用现今最常见的建筑技术和材料，寻求建筑空间与周围环境的完美融合。通过遮光、避光等传统的营造手段，减少外部环境的气候变化对室内温度的影响。采用自然的通风和采光，使本项目成为低能耗的生态型建筑。

高效开放的空间组团
6栋独立的建筑作为办公总部，高效有序的建筑布置形式。两个公共开放空间的周边设有供出租和出售的南北组团，这些不同用途的组团通过一条中央绿带有机联系，但又互相独立。

智能化、花园式的办公环境
引入各种智能化的防灾、管理和通信等系统，以确保24小时人性化的办公建筑环境。绿色空间结合智能化的办公环境，使人在享受科技带给他们的便捷、安全、高效的同时，也可以感受大自然的魅力。

设计手法
生态手法的导入
通透的建筑及自然采光将外部的绿色环境引入室内，节省了人工照明的能源成本。屋顶绿化有利于建筑的保温，垂直绿化将有效避免和遮挡阳光的直射，减少空调等能源的消耗。建筑底层局部架空，同时设置绿色中庭和可开启的天窗形成拔风效应，达到自然的通风和采光。设计师在屋顶设置可随阳光调节角度的太阳能采集器，充分利用自然资源。

有效的空间布局
建筑排列有序且富于变化，加强了面对主路的景观的通透性和办公工序所期望的序列感。中央绿地贯穿整个基地，与地上和地下空间相结合，使其更具有壮观的空间效应和情趣。南北组团围绕两个下沉的公共开放空间展开，为基地提供更多的辅助设施和人与人交流的平台。人车分流，以形成合理的交通动线。此外，项目设有安全舒适的中央绿色步行系统。

引入现代科技的绿色环境
导入IP电话、卫星设备等现代高效通信系统。设置较高的建筑层高，为将来各线缆的预埋留有余地。利用光电转换系统和太阳能集热器，对建筑供能进行补充。对废水和废物的处理，形成可用于冲厕和灌溉的"中水"及园区内绿化使用的肥料。绿色的休息中庭、阳台和户外连贯的生态步行系统、开放空间形成宜人舒适的办公环境。

EXALTIS TOWER AT LA DÉFENSE
拉德芳斯 Exaltis 大厦

Architects: Arquitectonica
Associate architects: Agence d'architecture Bridot Willerval
Location: Paris, France
Gross floor area: 23,000 m²
Height: 72.35 m
Photographer: Paul Maurer

设计机构：Arquitectonica
合作设计单位：Agence d'architecture Bridot Willerval
项目地点：法国巴黎市
总建筑面积：23 000 平方米
高度：72.35 米
摄影：Paul Maurer

Project Description
A new 15-story 23,000 m² office building and 3-level underground parking garage lie in the La Défense District in Paris. Conceived as a tower of glistening glass, Exaltis defines the axis of the avenue and replaces a grim gray viaduct. The tower is flanked on one side by a landscaped plaza and on the other by a linear park.
Along the avenue a monumental glass wall is modulated by the regular cadence of tall slender circular columns that march along the facade and act as oversized lanterns. Across from them along the core, two curved surfaces, one in frosted glass and the other in black granite, meet at a midpoint. This lobby stretches the entire length of the building. Grand stairs connect the two levels establishing a sculptural presence.

Design Intent
The linear site was at one time the center median of a major avenue which later was elevated and became a viaduct. The viaduct was removed to create the site but its history of speed and movement called for a dynamic form. There is nothing static about the site even today and the building form responds to this call.
The prism began its design life as a pure rectangle-functional floor plate, efficient core, ultimately flexible in its orthogonal plan, fitting snugly on the tight site. Energy was given to this otherwise static form by the manipulation of its end facades. Two curves splay from an imaginary point below and shoot towards the skies, giving the building a thrust towards the esplanade of La Défense.
Horizontal metal bands add a directional texture and depth to the semitransparent bottle-green skin. The building is at times solid, at times translucent.
This lobby is an expressionist synthesis of the forces that shapes the exterior. The multiply curves like a wild expressionist scribble on a sketchpad. They describe the emotion of architecture and the emotion of the architect.

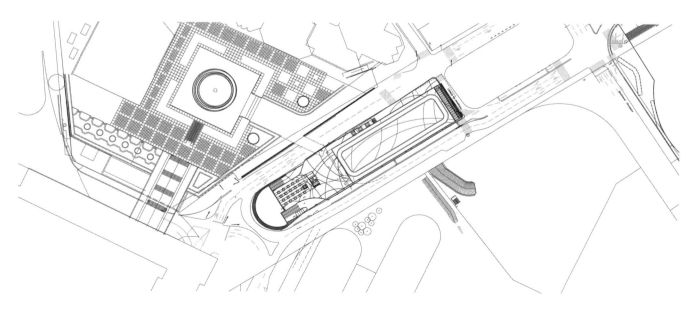

Masterplan 规划图

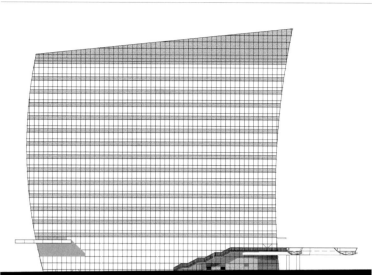

North Elevation　北立面图

East Elevation　东立面图

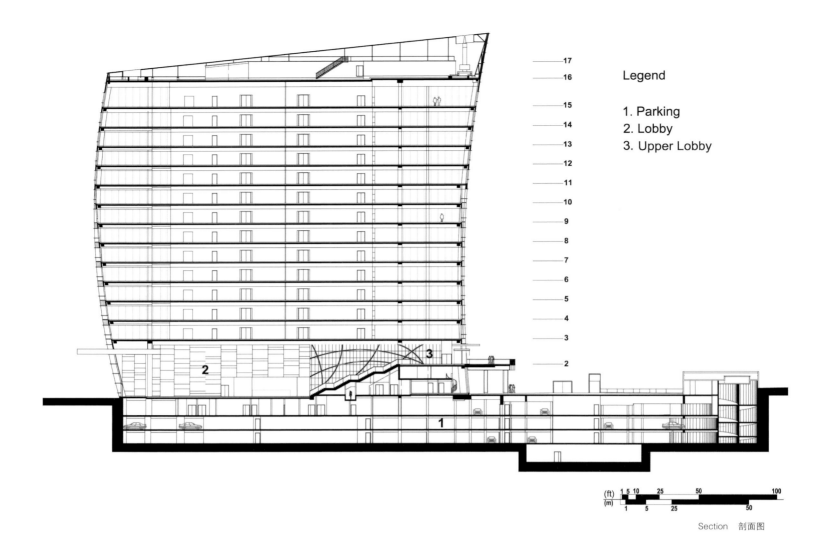

17
16
15
14
13
12
11
10
9
8
7
6
5
4
3
2

Legend

1. Parking
2. Lobby
3. Upper Lobby

(ft) 1 5 10   25   50   100
(m) 1   5   25   50

Section 剖面图

项目说明

这座占地面积23 000平方米、带有3层地下停车库的全新15层办公楼，位于巴黎的拉德芳斯区。以一座闪闪发光的玻璃大厦出现，Exaltis取代了死板的灰色高架桥，更好地诠释了大道轴线。大厦的一侧是一个景观广场，另一侧则是一个带状公园。

沿着大道，醒目的玻璃幕墙中间隔布置有细长的圆柱，圆柱依附幕墙游走，好似一个大型的灯笼。在它们对面，有两个沿着轴心的弯曲表面——一个是毛玻璃，一个是黑色花岗岩——交会于正中央。大堂延伸了整座建筑的长度。豪华的楼梯连接了富有雕刻效果的两个楼层。

设计意图

这个带状地块曾经是主大道的核心区段，后来被抬升成为一座高架桥。高架桥被移除后建造了这座大厦，然而高架桥速度与运动的过去，仍然需要大厦以一种流动形态表现。即使今天这个地块也毫无静态感，大厦外形不得不响应这一需求。

设计伊始，棱柱就作为一个纯粹的矩形功能性底板，高效核心，最终灵活地运用到它的直角面上，从而更好地适应这个繁华的地块。通过对其立面的处理，将能源传送到这个动态建筑中。两条曲线从下面假想的一个点出发，伸向天空，将建筑推向拉德芳斯的滨海大道。

水平金属带为深绿色半透明外表皮增添了一种方向性的纹理和深度。该建筑时而是实在的，时而是透明的。

大堂用表现主义手法塑造外部，表现力量的结合。多样的曲线宛如一个狂野的表现主义画家在画板涂鸦。它们以此描述建筑的动态和建筑师的情感。

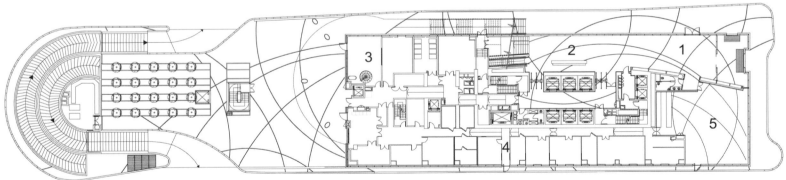

Plan 1 平面图 1

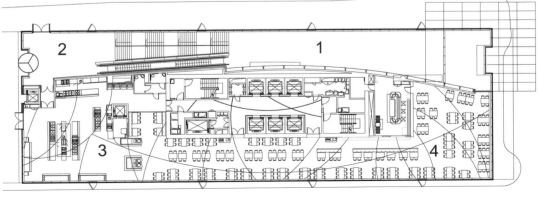

Plan 2 平面图 2

商业建筑　COMMERCIAL BUILDING

In the narrow sense, commercial buildings refer to public buildings used for exchange and circulation of commodities, as well as places where people are engaged in various business activities. Commercial buildings can be divided into retail stores where all kinds of daily necessities and means of production are sold, shopping malls and wholesale markets, trading places for financial securities, various service buildings such as hotels, restaurants, cultural entertainment facilities, shopping centers, commercial streets and clubs.

Modern commercial buildings emphasize not only the pursuit of commercial interest, but also social value, public interest, cultural taste and the impact on people's lifestyles. To estimate whether a commercial building is provided with aesthetic value, we not only see its external factors such as decoration materials, facade design, proportion and color, but also emphasize the participation sense of people. Modern society is a people-oriented society, and people are the main body of the society and space. To place people in an architectural building, the scale of the building should be considered carefully. Commercial buildings, especially large commercial building complex, should be complicated but not messy in form, bulky but not distorted, dignified but approachable, and its facade is also a symbol of commercial culture. The commercialization of architecture promotes the diversification of design approach in commercial buildings. After architects' endless quests for innovation, current architectural design methods are gradually maturing. Plenty of new building terms are generated to enrich people's lives and meet people's physical and spiritual demands.

Modern commercial buildings play an important role in urban public buildings, impacting the development, construction, image and environment of cities. They also play an important role in modern landscape, containing rich contents in people's daily lives, enriching the splendor of modern buildings through their unconventional, stylish and beautiful exterior designs. Thus modern commercial buildings have become the iconic symbol of modern civilization.

　　狭义上的商业建筑是指用来进行商品交换和商品流通的公共建筑，以及供人们从事各类经营活动的建筑物。商业建筑的主要类型有销售各类日常用品和生产资料等的零售店、商场和批发市场，金融证券等行业的交易场所，各类服务业建筑，如旅馆、餐馆、文化娱乐设施、购物中心、商业街和会所等。

　　现代商业建筑在追求商业利益的同时，也非常重视其自身的社会价值、公共利益、文化品位以及对人们生活模式带来的影响。判断一座商业建筑是否具有审美价值，不仅仅是看它的装修材料、立面设计、比例、色彩等外部因素，更重要的是强调人的参与意识。现代社会是一个以人为本的社会，人是社会的主体，也是空间的主体。建筑形体的尺度经过周密考虑，以使人能置身其中。商业建筑，尤其是大型商业综合体建筑的形体应该繁复而不凌乱、体量庞大而不失真，它既是"超人"的又是"亲人"的，其外立面也是商业文化的表征。建筑的商业化促使了商业建筑外观设计手法的多样化，各式各样的设计五花八门、层出不穷。经过设计师们的努力创新，当下的建筑设计手法正在逐步走向成熟，并且产生了很多建筑类的新词汇，更好地丰富了人们的生活，满足了人们物质和精神的需要。

　　现代商业建筑是城市中的重要公共建筑，它对城市的开发和建设，对城市面貌和环境的塑造产生了重要影响，在现代景观中具有显著的地位，涵盖了人们日常所需的丰富内容。标新立异、时尚美观的外形设计使商业建筑为现代建筑增添了无限风采，成为现代文明的标志性符号。

TAIPEI GALLERY
台北艺廊

Architects: P&T Group
Location: Taipei, Taiwan
Site area: 34, 200 m²

设计机构：巴马丹拿集团
项目地点：中国台湾台北市
总用地面积：34 200 平方米

This project is located in Xinyi District, whose design concept is to create a leisure shopping center full of sunshine and greenery, which will be different from other malls in Taipei Metropolitan Area. The wide oval skyline central plaza is surrounded by the three-storey-high podium, brand shop, shopping mall, and the high ceiling gallery which is easy for lighting on the first floor to get the main flow intersections connected. Combining with the delicate construction, water features and landscaping, it creates a pleasant, harmonious and friendly shopping environment. Two storeys of restaurant and owner-occupied space are laid out inside the tower in the north of the mall.

　　本案位于台北市信义区内，设计理念源于创造一座充满阳光、绿意盎然的休闲购物中心，以区别于现有台北都会区的其他商场。三层高的平台、品牌店、商场以及高挑空采光廊道沿着广阔的椭圆天幕中央广场布置，首层的灯光便于捕捉到主要的客流交会点。它与细致的建筑、水景及绿化互相配合点缀，创造出一个宜人、和谐、人性化的购物环境。商场北面的塔楼内设有两层餐厅层及两层业主自用空间。

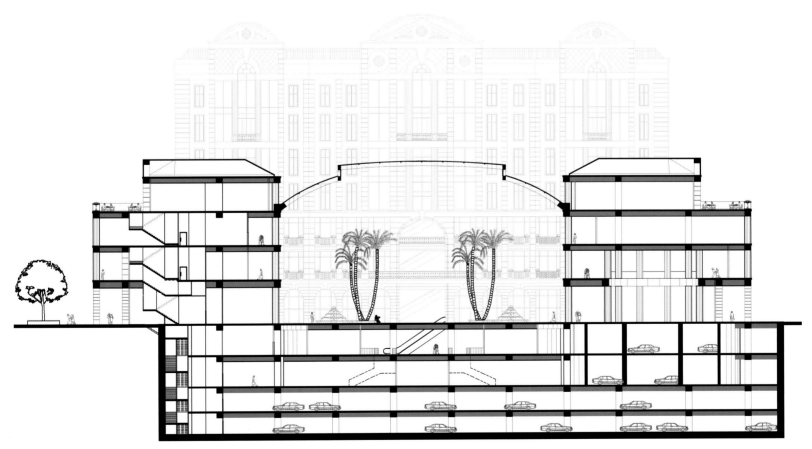

Section   剖面图

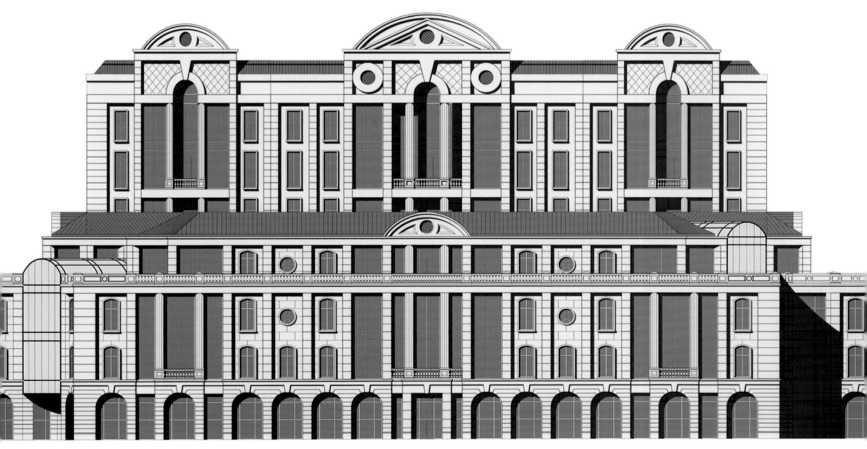

Elevation　立面图

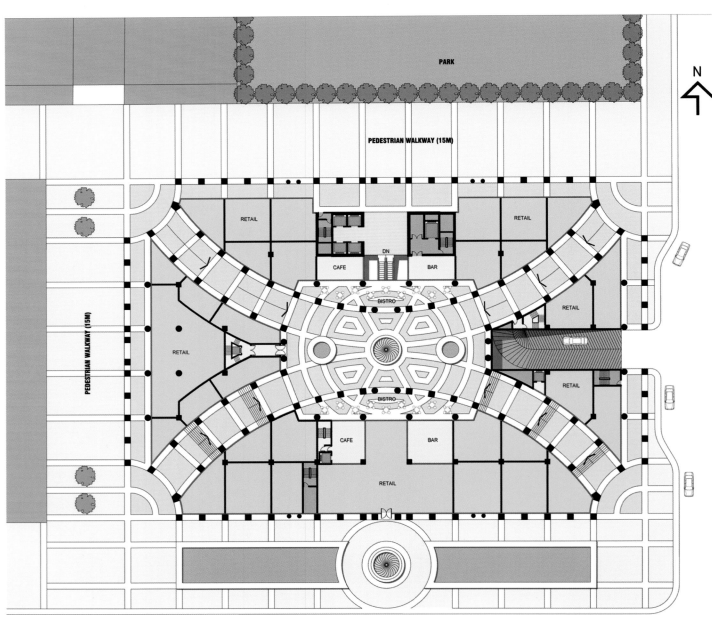

Plan 平面图

Plan 平面图

CROWNE PLAZA SHENYANG INTERNATIONAL
沈阳国际皇冠假日

Architects: MT Design & Engineering Co.td.
Location: Shenyang, China
Site area: 9,640 m²
Gross floor area: 49,087 m²

设计机构：香港美腾设计工程有限公司
项目地点：中国沈阳市
总用地面积：9 640 平方米
总建筑面积：49 087 平方米

Located in the South Huanghe Street, Huanggu District of Shenyang City, the project covers a total site area of 9,640 m². The Crowne Plaza is 18-storey-high with two basement floors. With 295-guest rooms including presidential suites and varied luxurious rooms, it is one of the biggest five-star international hotels in Shenyang.

Layout Design
Crowne Plaza Shenyang Parkview is composed of a main building and a podium which looks like a crystal cube with 45o. Inside there are 1,600 m² convention rooms, stylish restaurants, unique VIP lounge, business center, parking lots, fitness faculties, etc.

Architectural Design
Large-area colorful crystal sections are used in the lobby. The colorful crystal balls around the ceiling chandelier have played melodious music to make the space harmonious and show the elegance of the hotel.

The Chinese restaurant is designed with traditional Chinese aesthetics and philosophy. Combining with modern techniques, the space is designed in garden style to show the specific culture of Qing Dynasty.

The layering glass curtain wall has enhanced the sense of rhythm created by the lines. The hotel advocates the harmony between people and nature. Soft background music and the flowing water beside the tables have shown leisure and comfort to the most. The dining tables are designed in elegant style with polished wooden material which looks simple and modest. The lighting fixtures are made of natural materials and the vivid carved art works look smart and delicate. The columns and the beams of the cafeteria are specially designed to echo the surrounding environment.

Room is the place where people stay for most of the time. So great room atmosphere will make the hotel perfect. The design of the rooms pursues simplicity and fashion by combining bright background colors with soft light to make the guests feel comfortable. Long passageway foiled by carpet looks quiet, luxurious and comfortable. In addition with melodious background music, there is a feeling of "home".

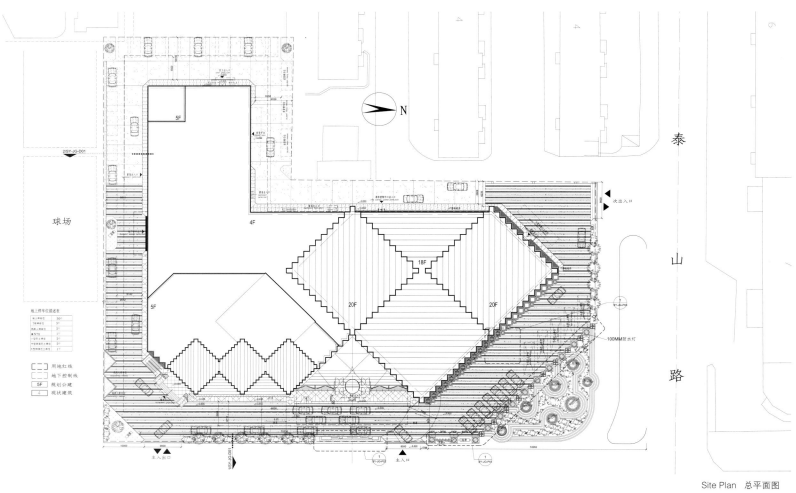

Site Plan 总平面图

　　项目位于沈阳市皇姑区黄河南大街，总用地面积 9 640 平方米。酒店地上 18 层，地下 2 层，拥有总统套房、各类豪华客房 295 间，是沈阳地区现阶段规模最大的一家国际五星级酒店。

　　规划布局
　　沈阳国际皇冠假日酒店由外观宛若 45° 旋转的水晶几何方体立方体主楼和裙楼组成，设有 1 600 平方米的会议厅、风格迥异的各种餐厅、独特的贵宾休息室、商务中心、停车场和健身设施等。

　　方案设计
　　大堂中空位置采用大面积的七彩水晶切面，造型美轮美奂。环绕天花吊灯的七彩水晶球奏出如风铃般清脆的乐章，令空间呈现出一片和谐气象，突显酒店的高贵典雅。
　　酒店的中餐厅设计从中国传统文化中吸取养分，融入中国传统美学、哲学等元素，结合现代设计手法，使用清代园林厅堂的空间概念，展现出清朝特有的文化风韵。
　　层叠的玻璃幕墙增加了建筑外观的线条节奏感。酒店崇尚人与自然的和谐统一。柔和的背景音乐、餐桌边潺潺的流水，将度假的闲逸之情表现得淋漓尽致。餐台造型简约，采用抛光的原木材料，简单又不失稳重。自然材质的灯具，生动形象的雕刻工艺品，透露着灵巧和精致。自助餐厅的柱和横梁有意而为，以此来呼应周围环境。
　　客房是客人在酒店内停留最久的空间，营造良好的客房气氛可让酒店的整体配套更加完美。客房整体设计追求简约时尚，明快的背景色彩与柔和的灯光，让客人倍感舒心。长长的客房通道，于地毯的陪衬下在静谧中流淌着华丽与舒适，加上悠扬的背景音乐，营造出"家"的感觉。

2nd Floor Plan　2层平面图

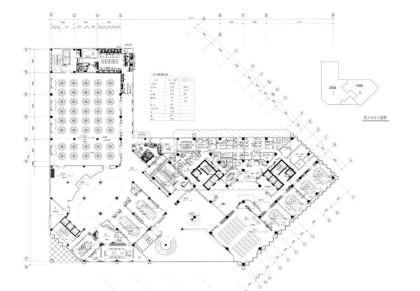

3rd Floor Plan　3层平面图

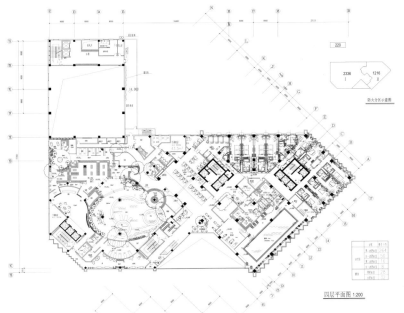

四层平面图 1:200

4th Floor Plan　4层平面图

ZHENGZHONG GOLF YINXIU SHANJU HOTEL,SHENZHEN

深圳正中高尔夫隐秀山居酒店

Architects: Xingtian Construction and engineering design firm, Shanghai
Location: Shenzhen, China
Site area: 41,507 m^2

设计机构：日兴设计·上海兴田建筑工程设计事务所
项目地点：中国深圳市
总用地面积：41 507 平方米

Zhengzhong Golf Yinxiu Shanju Hotel gathers diverse functions, including golf, entertainment, business, meetings, receptions etc. Located in the southwest of Zhengzhong Golf Club, Longgang District, next to Longhu and the golf course, the base is endowed with beautiful natural environment which creates the conditions for the quality of the hotel. Architects take the "earth" and "wood" as design concept, make architecture keep in harmony with environment, refine and blend geographical features, integrate architectures into environment with an appropriate scale so as to employ the elements of returning to nature through the design.

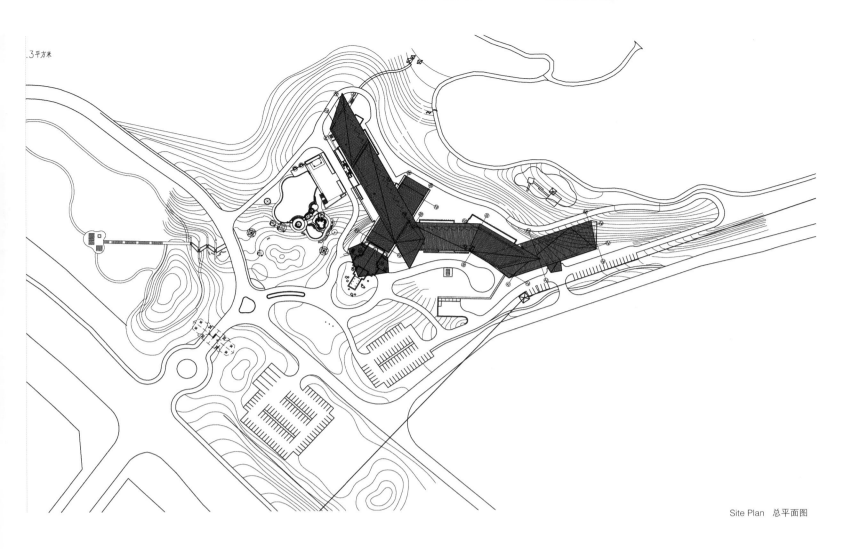

Site Plan　总平面图

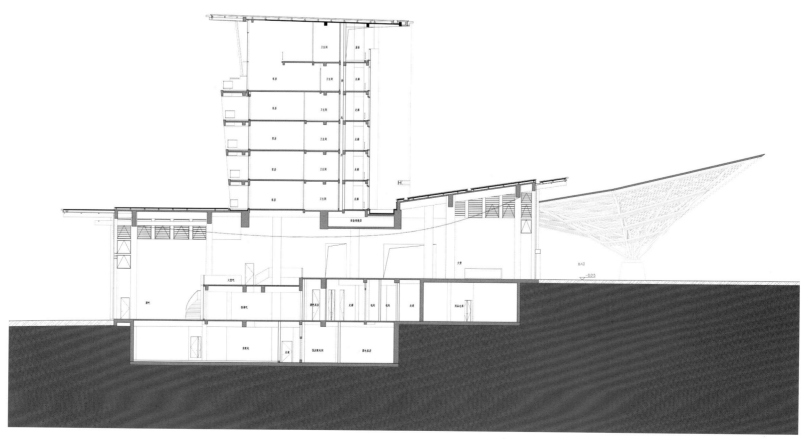

Section 剖面图

The main entrance of the hotel uses wooden materials to form a long-span space of 27 m-width along the vertical and horizontal direction and an axial symmetry pentagon. In this way, the building expands from center to the east and west sides along the fold lines rhythmically. All the first floor of the building is large spaces for public activities. There are meeting rooms, VIP reception room, business center in the west of the lobby, while shops, western restaurants and banquet halls etc, in the east. Above the second floor, it has 210 guest rooms including standard rooms, luxury suites, accessible rooms etc. Administrative reception hall stands on the top floor with rooms of open-plan spaces. In its east side, there is the Presidential Suite. In order to get a better view and vision for the rooms, the designers use top interlayer of the first floor to reach structure and equipment conversion. The mansard plane as well as the structure conversion make guest rooms and the main structure of the column grid form an angle and decrease the standard width of the column grid in guest rooms so as to unify the structure column wall. By creating a three-dimensional staggered architectural space, the architects perfectly solve the contradictions such as landscape, structure, equipment and space inversion of the building etc. to make room spaces more contracted. To get better natural ventilation and lighting, guest rooms are sited on the landscape side towards to the court. So guests can feel the great nature even when walking in the corridor. Architects put an outdoor bathtub designedly on the balcony towards to the golf course. Guests will be relaxed to enjoy the scenery of the golf course when bathing.

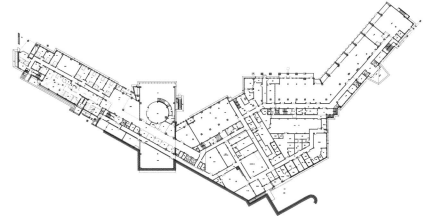

Underground 1st Floor Plan　地下 1 层平面图

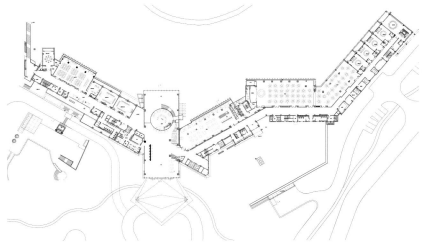

1st Floor Plan　1 层平面图

酒店东北侧夜景透视

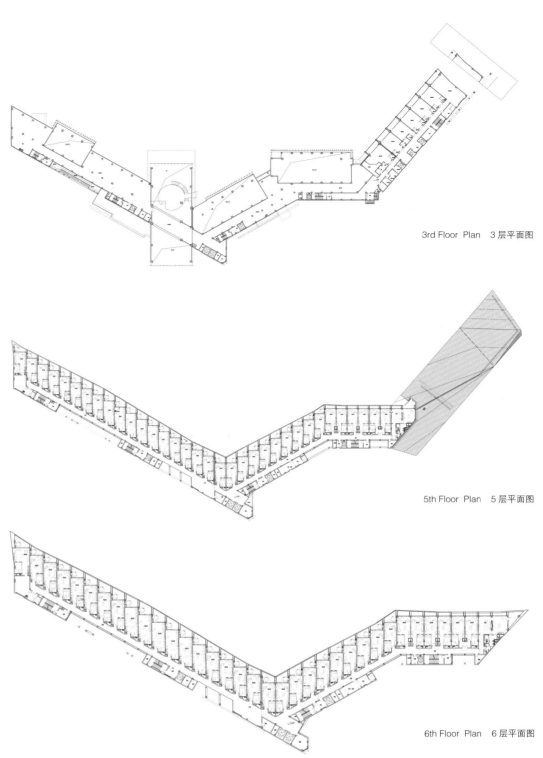

3rd Floor Plan　3层平面图

5th Floor Plan　5层平面图

6th Floor Plan　6层平面图

酒店入口日景透视

　　正中高尔夫隐秀山居酒店是集高尔夫运动、休闲娱乐、商务、会议、接待等功能于一体的休闲度假酒店，地处深圳市龙岗区正中高尔夫球会所西南侧，建筑紧邻龙湖和高尔夫球场，基地内自然环境优美，为酒店品质创造了条件。建筑师把"土""木"作为设计构思理念，以营造建筑与自然相和谐的环境为目标，提炼和融合地域特征，以得体的尺度把建筑嵌入环境中，将回归自然的元素始终贯穿于设计中。

　　酒店主入口采用木构材料沿纵横两个方向形成27米的大跨度空间，形成中轴对称的五边形，建筑也由此中心向东西两侧沿着折线有节奏地展开。建筑首层均为公共活动的大空间，大堂西侧设有会议室、VIP接待室、商务中心等，东侧主要设有商店、西餐厅、宴会厅等。二层以上主要为酒店客房，共有210间，分设有标准客房、豪华套间、无障碍客房等，顶层设有行政接待厅，其客房为跃层空间，最东侧是总统套房。为使客房获得更好的景观和视野，设计利用首层的顶部夹层进行结构和设备转换。折线形的平面加之结构的转换，使客房与主体结构柱网形成一个角度，并将客房的柱网间距减小，使结构柱墙一体化。建筑师通过建筑空间立体交错的方式完美地解决了景观、结构、设备和建筑的大小空间倒置等矛盾，使客房空间更加紧凑。为得到更好的自然通风和采光，向球场景观面单侧布置客房，宾客即使行走在走廊中也可感受到大自然的气息。在高尔夫球场一侧的客房阳台上特别设置了室外浴缸，宾客在沐浴时还可以欣赏高尔夫球场的景象，十分惬意。

MAOYE TIMES PLAZA
茂业时代广场

Architects: TJAD
Location: Shenzhen, China
Site area: 90,065.62 m²

设计机构：深圳市同济人建筑设计有限公司
项目地点：中国深圳市
总用地面积：90 065.62 平方米

Maoye Times Plaza was developed and constructed by Shenzhen Maoye Co., Ltd. The base is located in Nanshan District, facing the Wenxin Second Road on the east, Haide Second Road on the south, municipal road on the north, and Nanshan Bookstore on the north .The regular and distinctive site is located in the commercial and cultural center area, occupying new business portal of the coastal road, housing shops, leisure and entertainment facilities and culture resources. Building line requirements for each edge are not less than 8 meters. This site area is 10,926.48 m², with the functions of commercial space and office.

The design takes the status, feature of the site and surroundings into consideration to gain a careful planning, which unifies the economy and efficiency, improves the commercial and office quality. The sunken square, sky gardens and high-rise office are placed to keep good open visual corridors. The main entrance combines the sunken square with the leisure plaza to not only naturally separate the building from city roads, but also create a graceful surface. The traffic flow is rationally organized. The commercial spaces and office are independent relatively, sharing the main landscape of the base.

The building function is divided along vertical dimension, 1-8 levels for commercial podium, the 9th level for sky garden, the 10th level for overhead layer, the14th level for refuge layer, 11-25 levels for the office area. The architectural style as a landmark in the area is different from the general commercial building, expressing steady and solid social image.

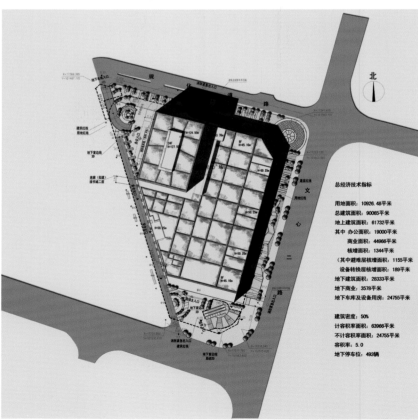

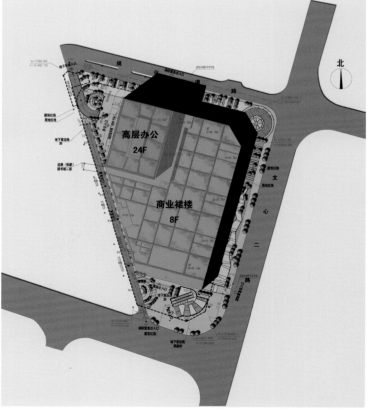

总经济技术指标

用地面积: 10926.48平米
总建筑面积: 90065平米
地上建筑面积: 61732平米
其中 办公面积: 19000平米
　　商业面积: 44966平米
　　核增面积: 1344平米
(其中避难层核增面积: 1155平米
设备转换层核增面积: 189平米)
地下建筑面积: 28333平米
地下商业: 3578平米
地下车库及设备用房: 24755平米

建筑密度: 50%
计容积率面积: 63966平米
不计容积率面积: 24755平米
容积率: 5.0
地下停车位: 493辆

高层办公
24F

商业裙楼
8F

Plan 1 平面图 1

Plan 2 平面图 2

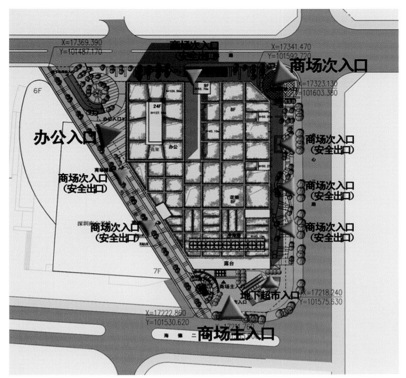

Plan 3 平面图 3

Plan 4 平面图 4

The unique facade of the building uses changeable vertical metal decorative strips to gain the sense of continuity and partition, steady and lively. The showcases made from stone and glass echoes with the upper vertical decorative strips contracture, combining with LED lamp to create glinting effect as a crown. The staggered order of vertical metal decorative lines and metal louvers make the building higher and straighter, leading a shocking sense.

The elevation adopts large area glass curtain wall. The LOW-E hollow glass of excellent transmittance and energy-saving are used to reduce the energy consumption of air conditionings and ventilation equipments, meeting the energy-saving requirements.

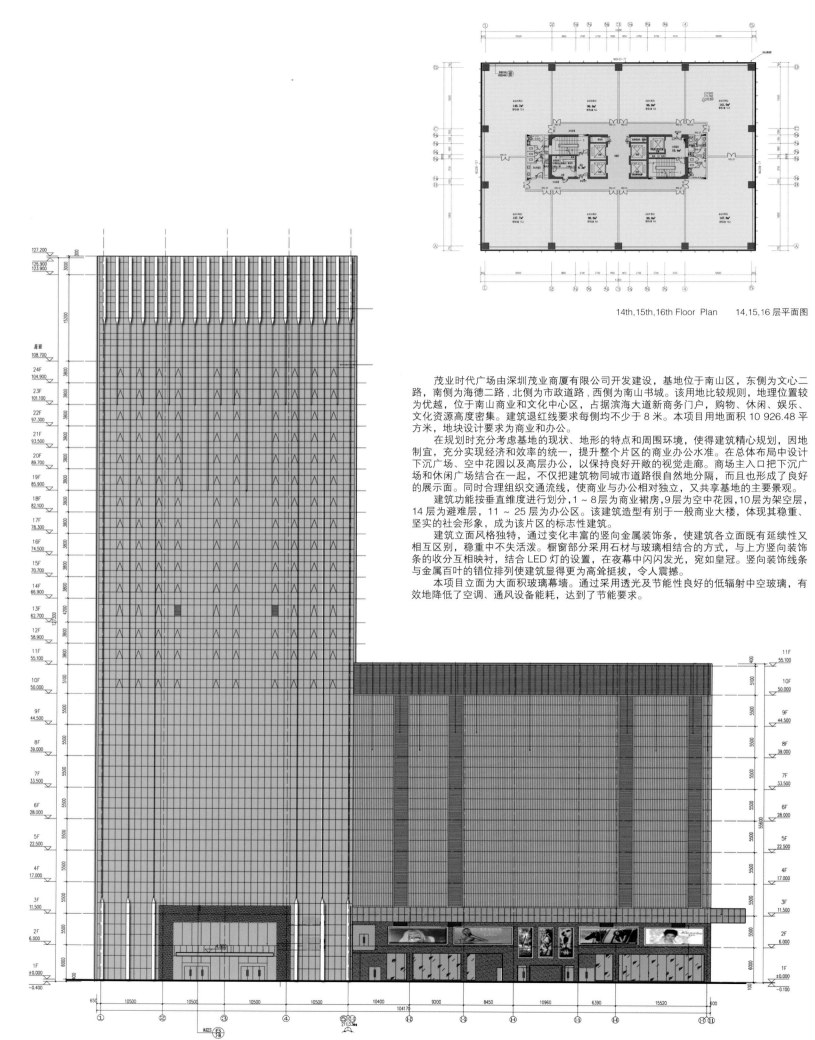

14th,15th,16th Floor Plan　14,15,16 层平面图

茂业时代广场由深圳茂业商厦有限公司开发建设，基地位于南山区，东侧为文心二路，南侧为海德二路，北侧为市政道路，西侧为南山书城。该用地比较规则，地理位置较为优越，位于南山商业和文化中心区，占据滨海大道新商务门户，购物、休闲、娱乐、文化资源高度密集。建筑退红线要求每侧均不少于 8 米。本项目用地面积 10 926.48 平方米，地块设计要求为商业和办公。

在规划时充分考虑基地的现状、地形的特点和周围环境，使得建筑精心规划，因地制宜，充分实现经济和效率的统一，提升整个片区的商业办公水准。在总体布局中设计下沉广场、空中花园以及高层办公，以保持良好开敞的视觉走廊。商场主入口把下沉广场和休闲广场结合在一起，不仅把建筑物同城市道路很自然地分隔，而且也形成了良好的展示面。同时合理组织交通流线，使商业与办公相对独立，又共享基地的主要景观。

建筑功能按垂直维度进行划分，1～8 层为商业裙房，9 层为空中花园，10 层为架空层，14 层为避难层，11～25 层为办公区。该建筑造型有别于一般商业大楼，体现其稳重、坚实的社会形象，成为该片区的标志性建筑。

建筑立面风格独特，通过变化丰富的竖向金属装饰条，使建筑各立面既有延续性又相互区别，稳重中不失活泼。橱窗部分采用石材与玻璃相结合的方式，与上方竖向装饰条的收分互相映衬，结合 LED 灯的设置，在夜幕中闪闪发光，宛如皇冠。竖向装饰线条与金属百叶的错位排列使建筑显得更为高耸挺拔，令人震撼。

本项目立面为大面积玻璃幕墙。通过采用透光及节能性良好的低辐射中空玻璃，有效地降低了空调、通风设备能耗，达到了节能要求。

Elevation　立面图

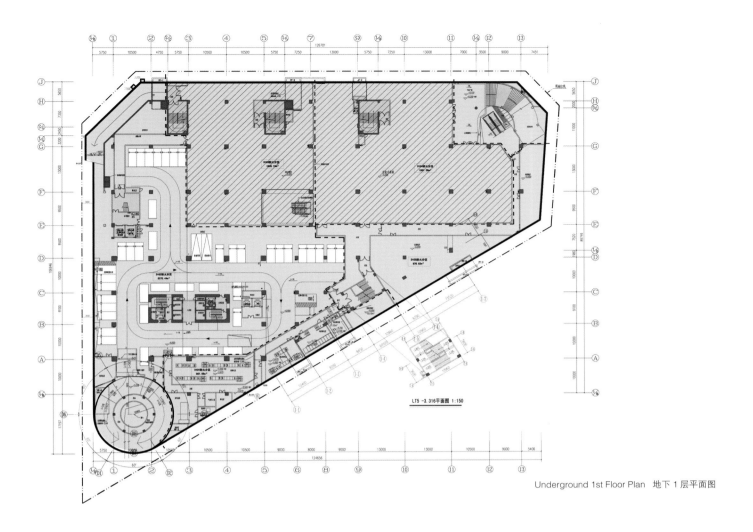

Underground 1st Floor Plan　地下 1 层平面图

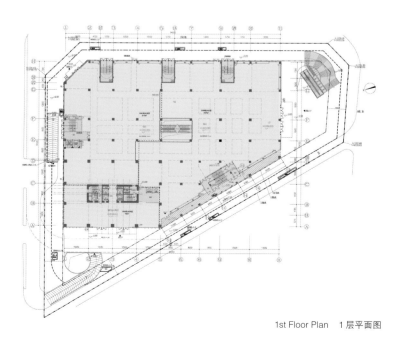

1st Floor Plan 1层平面图

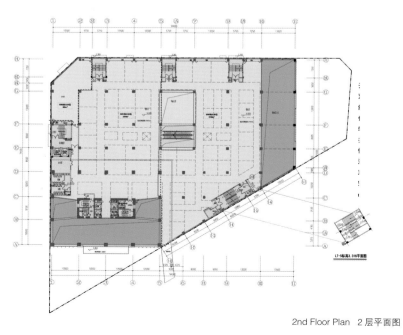

2nd Floor Plan 2层平面图

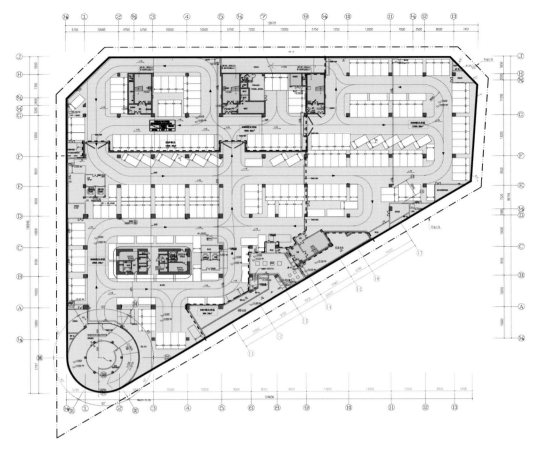

3rd Floor Underground Plan 地下 3 层平面图

GUANGZHOU SOUTH LAKE HOTEL
广州南湖宾馆

Architects: Sino-Sun Architects &Engineers Co.,Ltd.
Cooperative design: Guangzhou Newsdays Interior & Design
Construction Co.,Ltd.
Location: Guangzhou, China
Site area: 12,518 m²

设计机构：北京华太建筑设计工程有限责任公司
合作设计：广州集美组室内设计工程有限公司
项目地点：中国广州市
总用地面积：12 518 平方米

The project combines the modernism design principle and oriental traditional architecture to create an artistic conception of "oneness" partly through the observation, partly through intuition. The "oneness" of the South Lake Hotel is embodied in the oneness of China and the West, Heaven and Humanity, humanity and culture, humanity and nature. As to the oneness of China and the West, the design is modern which is mainly reflected on the process. As to the oneness of heaven and humanity, we have a special understanding for the "oriental" including spiritual dimension and regional characteristics. For spiritual dimension, we pay more attention to spatial size, the profile of the hotel and internal lines are full of oriental charm. For regional characteristics, it is full of spatial sense for it rises up from the bank of Nanhu Lake with unified and grand gesture. The harmony of Humanity and culture are reflected in the water. The existence of a large lake is amazing in Guangzhou downtown. As a kind of extension, the design produces a sense of belonging through various means, techniques and the perception, and creates culture atmosphere by using various artworks, soft contact media, and even voice, smell. For the oneness of humanity and nature, the design absolutely respects for environment and sustainability.

　　本项目采用现代主义的设计原则与东方传统相结合的手法，部分通过观察，部分通过直觉，以创造一种"合"的意境。南湖宾馆之"合"，主要体现为中西方之合，天人之"合"、人文之"合"、人与自然之"合"。在中西方之"合"上，我们的设计是现代的，一种经过传统过滤的现代感，主要体现在工艺上。而天人之"合"，在于我们对"东方"有着特殊的理解，其中包含精神层面和地域特性。从精神层面上说，空间尺度是我们"斤斤计较"的，酒店的轮廓以及室内的线条都充满东方韵味。从地域特性来说，空间感十足，它立于南湖之畔，气韵统一而气势浑然。人文之"合"则体现在水上，在广州市区，拥有一大片湖水是让人惊叹的。作为文脉的延伸，通过各种手法、技术和感知来建造一种归属感，加上各种艺术品，软性接触媒介，甚至声音、嗅觉组成文化的氛围。人与自然的"合"上，无论基于对令人惊叹的基地的考量，还是为节省成本以及求得与环境的统一，设计始终秉承对环境绝对尊重的原则。各种技术工艺的使用都是为可持续发展而设计。

建筑

1、湖畔楼功能布局基本保持原样，集餐饮、娱乐服务、客房为一体。配以以坐拥南湖美景的无边泳池为中心的内庭院，形成一个广东省一流的特色精品酒店。

2、原建筑物现状外立面过于生硬、呆板且陈旧，为将其装饰营造出温暖、自然、亲切与环境有机统一的度假氛围，拟增加一宽600mm的装饰性阳台，采用各种现代材料如钢、木、清玻璃等进行有机地组织与搭配，形成一种既现代、清新同时又具休闲、轻松味道的立面形象，配以南湖美丽的湖景，能给人一种耳目一新的感觉。

3、增加连廊通往四、五号楼，使三幢楼联系紧密且人车分流，营造优良的经营环境。

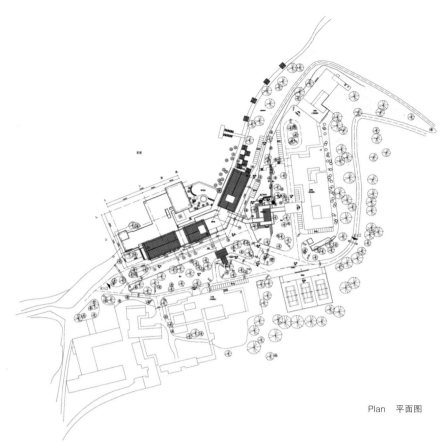

Plan 平面图

首层 | 湖景西餐厅

西餐厅位于优美南湖湖畔，景色迷人。如何使这种美丽的风景与室内空间融为一体，并得到升华是这次设计的主题。

设计中运用了可使空间自由分割的木结构屏风作为界面，以此在室内空间中勾勒出南湖山水的优美景色，使客人仿如置身自然中就餐。整个西餐厅成了自然与建筑有机联系的真正灰空间。

Plan 1 平面图 1

二层大堂设计风格现代、自然、简洁。利用线条的水平量体与垂直元素呈现实虚的空间效果，再透过大量的玻璃使用与间隔的安排，使自然景致随任何一种活动进行时，皆可一览无遗、仿如置身其中。

二层 | 大堂

Plan 2 平面图 2

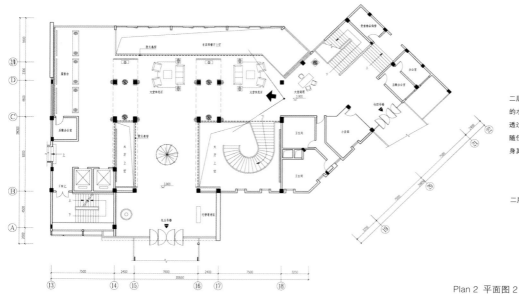

首层 | SPA

湖畔楼SPA区以简洁的建筑语言，点、线、面作为空间元素所构成。SPA豪华房以墙体分隔空间，二次光源使空间气氛更加突出，顾客在SPA的过程中更加舒适自在。立面以石材及布料作为主材，令空间更为洁净。SPA房入口用木地板与室外的露台相呼应，使空间从内到外得以连接，顾客更能感受室外的景致。

沐足区利用活动屏风分隔座位，灵活而又增添空间的灵活性。

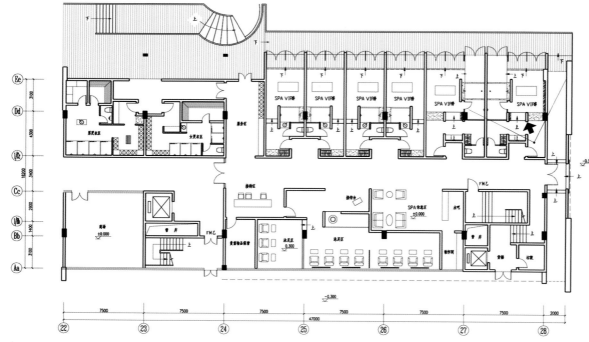

Plan 3 平面图 3

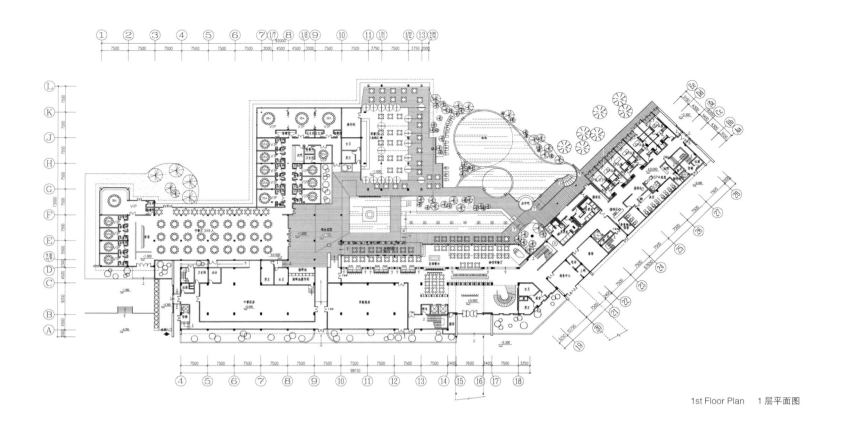

1st Floor Plan   1层平面图

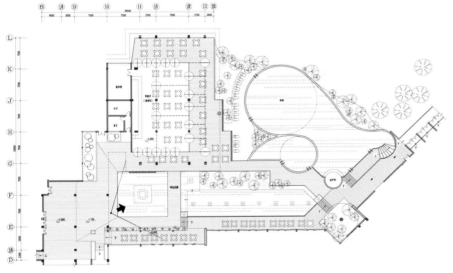

首层 | **邻水花园**

游泳池的设计以自然融合一体的概念使楼板水平线与湖面水平线相辉映，让泳池水位高于池缘，使置身泳池时，产生水、湖、天连成一体的奇妙感受。

Plan 1　平面图 1

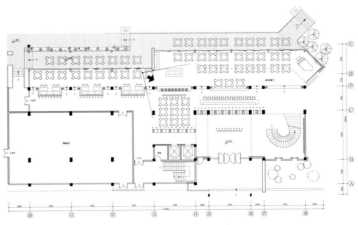

首层 | **大堂西餐厅**

东方设计注重于精神场所的营造，西方则追求对简约生活空间的营造，延续酒店的风格定位，西餐厅需要体现的正是一种东西方设计的营合。于是我们尝试以传统的木屏风与现代的设计手法相结合，将整个首层大堂的西餐部分与公共部分分割开，揭示了空间的虚实，利用自然元素满足了人的观赏心理，形成自然过渡。

Plan 2　平面图 2

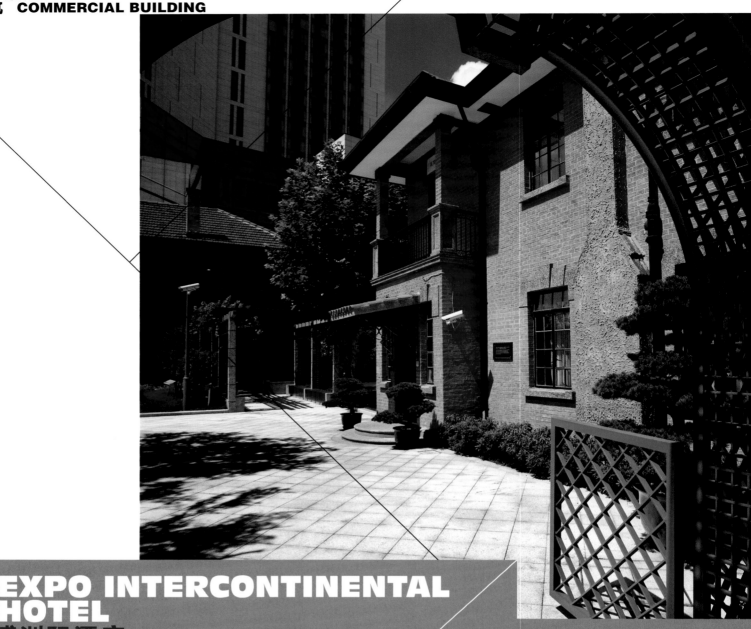

EXPO INTERCONTINENTAL HOTEL
世博洲际酒店

Architects: ECADI
Location: Shanghai, China
Site area: 24,632 m²
Gross floor area: 70,529 m²

设计机构：华东建筑设计研究院
项目地点：中国上海市
总用地面积：24 632 平方米
总建筑面积：70 529 平方米

The design of Expo Intercontinental Hotel respects to the master plan and design principle, focuses on the harmonious entirety of the city space order, avoids the visual impact from the contrasting elements and forms to produce the rejection by city spaces and master plan. In the master plan, the architects pay attention to the combination of new and old buildings that fuses the protected buildings into the hotel public areas. The traffic organization in the base has arranged the pedestrians, goods and logistics flows without mutual interference. As to the functions of the hotel, the Expo features are fully considered. The meeting function scale is enlarged. The building facade uses rational materials to strengthen the effect. The folded-plate-shaped main body is treated in two styles, conjunction of stone and glass as well as full glass. The windows of rooms are divided by same modulus. The up-and-down movement strengthens the facade surface to form the basic element of the building facade, giving a tall and strong image and exquisite details. The hotel fully expresses the Expo theme, "better city, better life", by using many green environmental and energy-saving technologies.

世博洲际酒店从城市设计的角度，尊重世博村总体规划与设计原则，注重城市空间秩序的整体和谐，避免对比强烈的元素和形体而产生视觉冲击，造成对城市空间的排斥和对总体规划的离异。在总体设计中，注重新老建筑整体设计，使保护建筑完全融入酒店公共区。基地内交通组织梳理了客、货及后勤不同流线，避免相互干扰。酒店功能配置充分考虑世博会使用特点，扩大会议功能规模。建筑立面设计合理运用材料，以其不同质感加强效果的落实。主楼折板形体分石材和玻璃组合以及全玻璃两种不同处理。客房开启窗以相同模数划分，上下错动突出幕墙表面，形成了立面的基本元素，使建筑挺拔有力，细部精美。酒店运用多项技术创新，在绿色环保节能方面，实施多项技术措施，成为设计的绿色亮点，充分演绎"城市，让生活更美好"的世博主题。

Plan 平面图

Master plan 规划图

CHINA HUASHUIWAN RESORT AND HOTSPRING SPA
中国花水湾度假酒店及温泉水会

Architects: OAD Office for Architecture & Design
Construction drawing design: Chengdu Architectural Design & Research Institute
Client: China Railway Erju Baden Baden Hotspring Development Co., Ltd.
Location: Chengdu
Site area: 89,346.484 m²

设计机构：OAD 建筑设计事务所
施工图设计：成都市建筑设计院
业主：中铁二局巴登巴登温泉开发有限公司
项目地点：中国成都市
总用地面积：89 346.484 平方米

Project Introduction
Huashuiwan Celebrity Resort, located in Huashuiwan, a small town of Dayi County in national tourism resort area with an area of nearly 90,000 m², is a national five-star standard resort invested by China Railway Erju Co., Ltd.
The design proposal carries out layout by fully respecting local natural environment and cultural context, has strong holiday atmosphere full of flowers and birds. With the overall length of 282 m, the Resort has beautiful landscape and broad view. The whole structure of the Resort comprises three regions: landscape lobby, guestroom and international convention center. There are more than 200 suites. The international convention center can accommodate 600 people at one time.

Design Concept
Organic Life in the Mountain
The overall frame of the building takes the mountain as the main body, the plane is configured on the natural terrain of the base, the elevation combines outline of the background mountain and gives people a feeling that the building is growing out of the mountain.

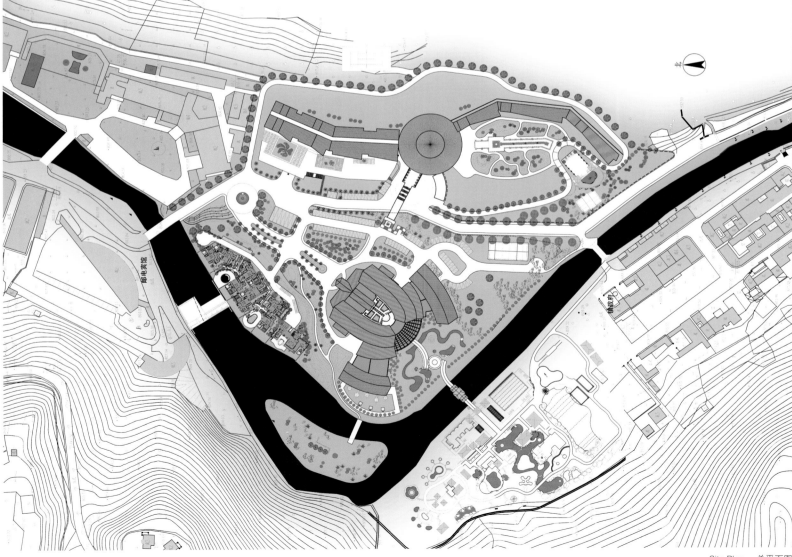

Site Plan   总平面图

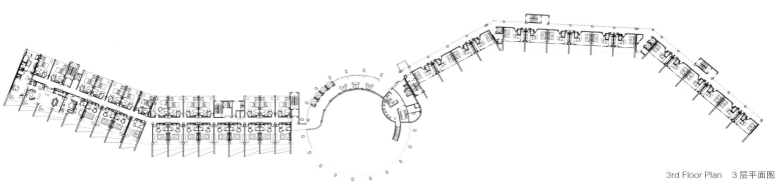

3rd Floor Plan　3层平面图

Indigenous Overlong Scale

The building design bravely adopts unconventional long scale to cater to continuous natural picture of the background mountain, thereby showing an excited visual angle rich of visual impact. The hotspring SPA is integrated with the circular hotel lobby, thereby reducing abrupt feeling brought by overlong scale of the main resort.

Inheriting the beauty of Sichuan dwellings

The building style design adopts Sichuan traditional swellings as the prototype, explores elegance of traditional roof elements and applies in the project with the adoption of modern engineering measures. Meanwhile, the building main body show strong Sichuan style and having high identifiability.

Staggered Change by Twisting on Mountain

The overall layout is to build on the momentum just as traditional hanging house. Rectangular template of the building can be twisted and deformed according to change of the mountain terrain, so as to form multiple transitional staggered aesthetic feeling.

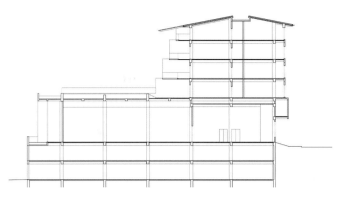

Elevation 1　立面图 1

项目简介

花水湾名人度假酒店地处国家级旅游度假区大邑县花水湾小镇，占地近 9 万平方米，是由中铁二局斥资打造的国家五星级标准主题度假酒店。

设计方案充分尊重当地的自然环境与人文脉络进行布局，具有浓郁的和花鸟共语的度假氛围。酒店全长 282 米，景观优美，视野开阔。酒店整体建筑由景观大堂、客房、国际会议中心三个区域构成，拥有 200 余间套房。国际会议中心能够同时接待 600 人的会议。

设计构思

生于山中的有机生命

建筑整体形态构架以山为母体，平面依托基地的自然地形布置，立面结合背景山体的轮廓线起伏，给人一种建筑生于山中的感觉。

因地制宜的超长尺度

建筑设计上大胆地采用了超常规的尺度迎合背景山体连绵的自然图面，呈现出一个极具视觉冲击力的震撼视角。同时温泉水会与圆形酒店大堂呼应成为一个整体，减少了主体酒店超长的尺度带来的突兀感。

传承四川民居之美

建筑造型设计以四川传统民居为原型，充分发掘传统屋顶元素的秀美，并以现代工程手段应用到项目中。同时建筑整体材质呈现出了浓郁的巴蜀风情，具有显著的可识别性。

依山扭曲的错落变化

整体布局参照传统吊脚楼的依山就势而建，层叠进退。根据山体地势的变化，将建筑的长方体模板进行扭曲和变形，形成多道转折的错落美感。

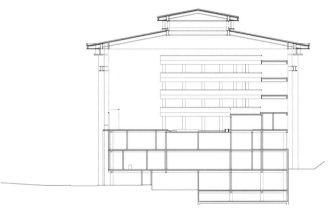

Elevation 2　立面图 2

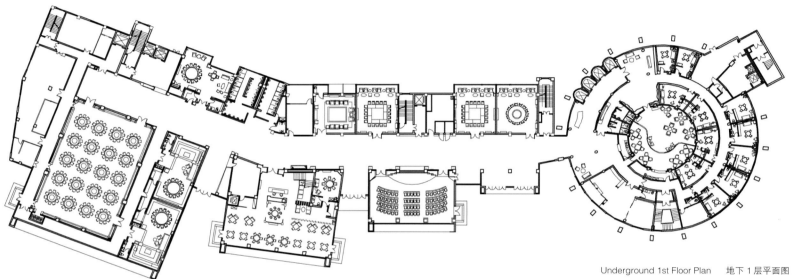

Underground 1st Floor Plan   地下 1 层平面图

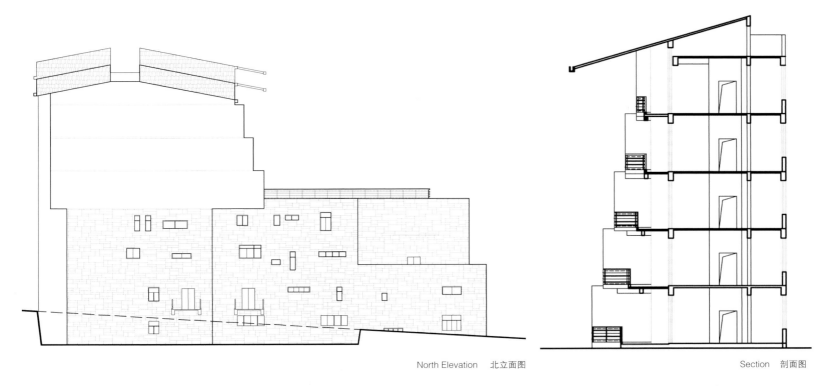

North Elevation　北立面图

Section　剖面图

TAIKOO HUI MIXED-USE DEVELOPMENT
太古汇

**Architects: Arquitectonica
Designer: Bernardo Fort-Brescia
Location: Guangzhou, Guangdong, China
Site area: 48,954 m²
Photographer: Taikoo Hui**

设计机构：Arquitectonica 建筑事务所
设计师：Bernardo Fort – Brescia
项目地点：中国 广州市
总用地面积：48 954 平方米
摄影：太古汇

Project Description

Tai Koo Hui is a landmark mixed-use development in Tianhe District, Guangzhou, China. The site is situated on a prime location and is measured approximately 48,954 m². The program includes: a 4-storey retail centre (120,000 m²); a 40-storey storey Grade "A" office tower (100,000 m²); a 28-storey Grade "A" office tower (60,000 m²); a 5-star, 28-storey, 286-key Mandarin Oriental Hotel (65,000 m²); a cultural centre (60,000 m²) with 1,000-seat theater, public library and performing arts center; associated ancillary facilities for car parking (780 cars); loading and unloading docks (42 nos.); mechanical and electrical plant and equipment; and related management and service facilities. The project will be connected by subterranean pedestrian tunnels to the Metro 1 underground train line station at Sports Road East and the Metro 3 underground train line station located on Tianhe Road.

Vision

The site is organized around the fundamental principle of maximizing open space experiences for the users of the complex and the community at large.

The podium setting for the development provides a new man-made geography from which the proposed buildings rise and under which functions thrive. The podium may be considered as a hill town with an active life within and below its surface.

Projecting skywards from the terraced park are two office towers and a hotel tower. Their crystalline forms have softly chamfered edges that provide more light into the heart of the development and give increased views outwards to the city. The towers are placed in a picturesque arrangement, at angles to each other.

太古汇位于中国广州天河区，是一个具有里程碑意义的多功能发展项目。太古汇地理位置优越，占地大约 48 954 平方米。该项目包括：4 层高的零售中心（室内总面积 120 000 平方米）；40 层的 A 级办公楼（室内总面积 100 000 平方米）；28 层的 A 级办公楼（室内总面积 60 000 平方米）；五星级的东方文化酒店，共 28 层 286 套房间（总面积为 65 000 平方米）；文化中心有一个可以容纳 1 000 人的剧院（总面积为 60 000 平方米）；公共图书馆和演艺中心；停车场相关的配套设施（780 辆车）；装载和卸货码头（42 号）；机械和电气厂房及相关设备；相关的管理和服务设施。该项目将通过地下行人隧道，连接体育东路的地铁一号线和天河路的地铁 3 号线。

选址是以用户和社区的最大利益为基本原则的。它以三维模式为出发点，以不同海拔的开放空间去效仿这个地区的自然地形。

为项目设置的平台从拟建的建筑物和功能繁荣方面提供了一种新的人造地理。这个平台为人们提供了一种积极的生活态度。

从沿着斜坡建造的公园里可以看到两座办公大楼和一座酒店大楼。它们水晶般的形状极富特色，为城市后续发展提供更多的思考空间，并且为城市提供更多的新意。大楼在相互选址安排上也给人一种如画般的感觉。

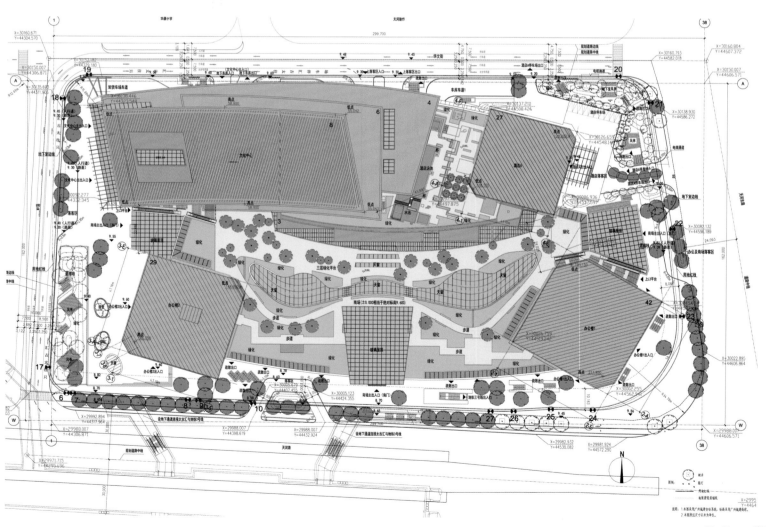

Site Plan 总平面图

BEGONIA BAY INTERCONTINENTAL RESORT HOTEL
海棠湾洲际度假酒店

Architects: Architects NIKKO · Shanghai Xingtian Architectural Design Group
Location: Sanya, China
Site area: 61,200 m²

设计机构：日兴设计·上海兴田建筑工程设计事务所
项目地点：中国三亚市
总用地面积：61 200 平方米

Begonia Bay is a pearl of vacationland in Sanya, China. Begonia Bay, Yalong Bay, Donghai Bay, Sanya Bay and Yazhou Bay are five famous bays in Sanya. They are treasure places, possessing exquisite sceneries. The project is located in the waterfront hotel band of so-called "national coast", Begonia Bay, 28 kilometers from the downtown of Sanya. Design perfectly fuses the beautiful natural environment, natural and organic architecture and unique architectural style together to create a iconic resort hotel for Begonia Bay. The design was inspired by the sea wave and shell, combining the soft feature of the wave and the unique form of shell. The hotel's overall layout takes terrain and sea view into consideration, stretches as U-shape, makes full use of the relative elevation and coastal landscape, optimizes the landscape orientation and maximizes the view for hotel rooms, the lobby of the hotel and restaurants. The step back layer by layer achieves the zero distances contact between visitors and sea view and provides the fresh air and subtropical scenery for the users.

　　海棠湾，作为度假胜地中国三亚市的一颗明珠，与亚龙湾、大东海湾、三亚湾、崖州湾并称三亚五大名湾，其风光旖旎，是受到造物主恩宠的风水宝地。本项目坐落于有国家海岸之称的海棠湾一线滨海酒店带，距三亚市中心 28 千米。设计将优美的自然环境、自然有机的建筑形态和独特的建筑造型完美融合在一起，打造了海棠湾标志性的洲际度假酒店。以海水波纹和贝壳作为构思创作的灵感来源，将海水波纹的柔和与贝壳形态的独特融合在一起。酒店的整体布局综合考虑了地形和海景视线两方面因素，呈 "U" 字形状展开，充分利用地形高差和滨海景观，尽可能地使酒店客房与酒店大堂、餐厅等拥有更好的景观朝向，实现景观视野最大化。层层后退的酒店客房，更实现了宾客与海景的零距离接触，享受了亚热带风光的清新与舒适。

Underground 1st Floor Plan 地下 1 层平面图

Underground 2nd Floor Plan 地下 2 层平面图

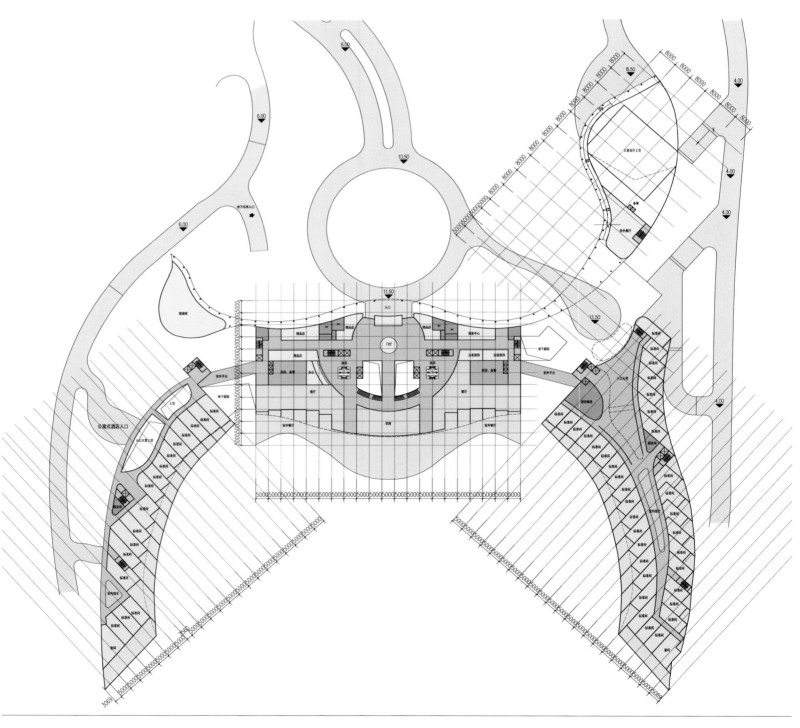

1st Floor Plan 1层平面图

Elevation 1 立面图 1

Elevation 2 立面图 2

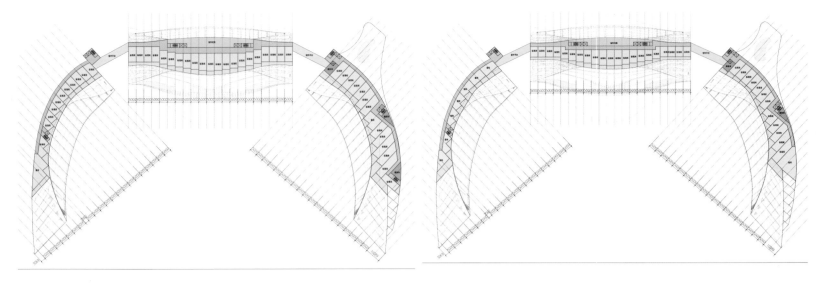

6th Floor Plan　6层平面图　　　　　　　　　　　　　　　7th Floor Plan　7层平面图

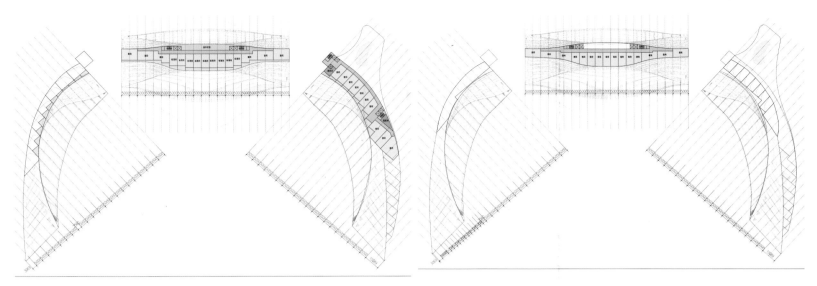

8th Floor Plan　8层平面图

9th Floor Plan　9层平面图

商业建筑   COMMERCIAL BUILDING

JINAN HANG LUNG PLAZA
济南恒隆广场

Architects: P & T Group
Location: Jinan, China
Site area: 256,263 m²
Photographer: H.G. Esch Photography

设计机构：巴马丹拿集团
项目地点：中国济南市
总用地面积：256 263 平方米
摄影：H. G. Esch Photography

Design Statement:
Parc 66 is a prominent new landmark inspired by the natural spring that flows through the city of Jinan. The two million square feet shopping mall overlooks Quancheng Plaza and connects the city's most popular open space to the historical Furong Street and the famous Daming Lake, creating a seamless connection to between these famous sites. The undulating form of the glass atrium glows at night and creates an enticing silhouette during the day forming an important backdrop for Quancheng Plaza. Parc 66 establishes a new urban landmark in the city of Jinan and transforms the cityscape into a fusion of old and new.

济南恒隆广场（Parc 66）这一显著的新地标建筑的设计灵感源于济南市内流动的天然泉水。占地两百多万平方米的商业中心不仅可以俯瞰到市内最热闹的泉城广场，还连接了历史悠久的芙蓉街和有名的大明湖，使这些著名景点自然地衔接。在建筑设计上，起伏的玻璃中庭在夜间可以发光，白天则展现出诱人的轮廓，成为泉城广场的主要背景。恒隆广场不仅为济南市创造了崭新的都市坐标，而且将市内著名的新旧景点互相融合。

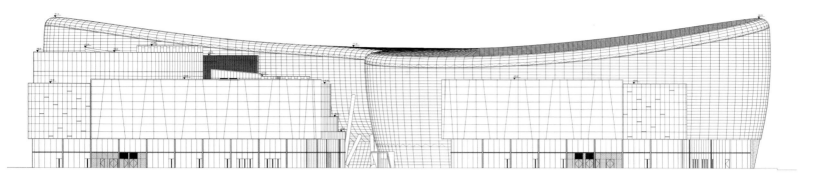

Elevation 立面图

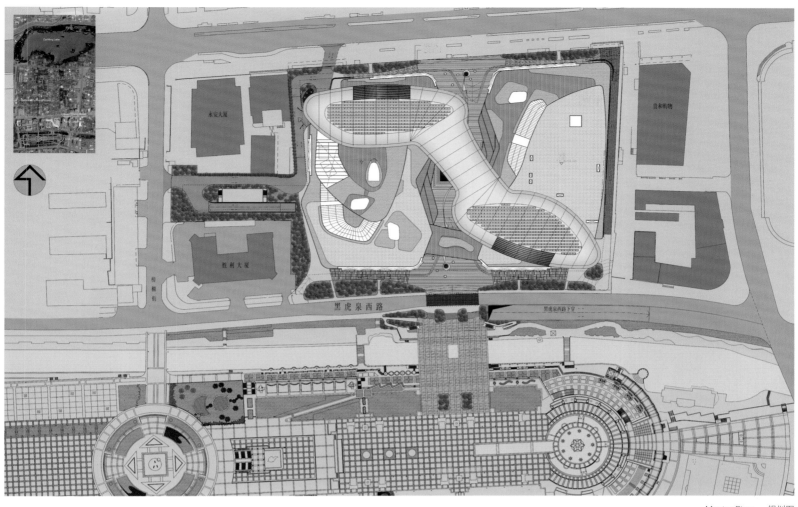

Master Plan　规划图

文化建筑　CULTURAL BUILDING

Being a kind of human behavior and way of life, culture is the bearer of the fruit of human civilization as well as an embodiment of the movement of civilization itself. Urban culture generally includes three parts: the material culture which represents the sensatory recognition of the city image, the technology culture which represents the spiritual identification of the city, and the norm culture which represents the regulatory distinction. It is the spirit and soul of the city. Any public cultural building can be considered as the material carrier of the spirit of the city, and the most important way to express its urban culture. However, the expression of architecture is showed by its model and space, and is melted into the architectural environment through the perception of historic culture and traditions.

When designers design public cultural buildings, they should integrate a variety of complicated factors together into a coherent whole and make a appropriate choice on the premise of comprehensive judgment. After the beforehand studies and summaries of design work, designers can begin with three aspects. Firstly, start with theme.This case usually aims at cultural displays and memorial sites for a specific topic and is always provided with relatively specific cultural performances and clear psychological anticipation on scale. Secondly, start with site environment. Many cultural buildings' cultural connotations are not provided beforehand, but depend on particular circumstances. In this case designers should proceed from the specific nature and human environments of the projects to extract useful information and manifest the true cultural connotations of buildings. Thirdly, start with contemporary values and aesthetics. With the continual appearance of new structures, new materials, new technologies, contemporary architectural forms are provided with protean manifestations. The emphasis of eco–energy saving, sustainability, merging and coexistence of public buildings and urban life, as well as sophisticated and variable building skins and rich design methods of contemporary public cultural buildings　are the most distinctive brand of the age for contemporary public cultural buildings.

Culture not only provides architecture with vitality, but also is spread widely by the expression of architecture. When building is provided with a cultural connotation, it is no longer a reinforced concrete pouring, but a cultural symbol of a city or a country.

　　文化是人类的一种行为方式和生存方式，是人类文明成果的承载，也是文明本体运动的体现。城市文化一般是由代表城市形象感观识别的物质文化、代表城市精神识别的技术文化以及代表规范识别的规范文化等组成，它是一个城市的精神与灵魂。任何一座公共文化建筑，都是城市精神的物质载体，是城市文化最重要的表达方式。而建筑的这种表达是通过造型和空间表现出来的，并通过对历史文化和传统的感悟形成认识，将之融入建筑环境之中。

　　设计师在设计公共文化建筑的时候，需要统筹考虑各种复杂因素，并在综合判断的基础上做出适宜的选择。经过对前期工作的设计研究和汇总，设计师可以从三个方面来切入：首先，以主题世界为切入点，这主要是针对某一具体的主题实践的文化展示和纪念场所而言，在这种情况下，公共建筑通常会拥有比较明确的文化性表现，对尺度也有明确的心理预期；其次，以场地环境为切入点，很多公共文化建筑所要呈现的文化内涵都不是事先给定的，这需要设计者从项目自身的具体情况出发，根据场地的自然环境和人文环境状况，提炼出有用的信息，表达出建筑所应有的文化内涵；第三，以当代价值观及审美观为切入点，伴随着新结构、新材料、新技术的不断涌现，当代建筑形式有了千变万化的表现，而强调建筑的生态节能和可持续发展，公共建筑与城市生活的共生融合以及复杂多变的建筑表皮，丰富多彩的创作设计，都为当代公共文化建筑烙下了最鲜明的时代烙印。

　　文化不仅赋予建筑以生命力，同时也通过建筑的表达而得到了加深与传播。建筑有了文化内涵，就不再是单纯的钢筋混凝土浇筑物，而是一个城市乃至一个国家的文化性标志。

MERCEDES-BENZ ARENA
梅赛德斯奔驰文化中心

Architects: ECADI
Construction: The Shanghai world expo performing arts center co., LTD
Location: Shanghai, China
Site area: 67,242.6m^2
Gross floor area: 140,277 m^2

设计机构：华东建筑设计研究院
建设单位：上海世博演艺中心有限公司
项目地点：中国上海市
总用地面积：67 242.6 平方米
总建筑面积：140 277 平方米

As one of the most important permanent pavilions for the World Expo, the Shanghai World Expo Cultural Center undertook various large-scale performances and activities to meet the large-scale performance requirements during the Expo. At the same time, the design took the follow-up utilization and sustainable development into consideration.

The main part of the performance center is multi-functional space with 18,000 seats, which is divided into many spaces with different scales and styles by flexible separation technologies to meet the demands for large and medium-sized arts performances, sports events, gatherings and celebrations etc. The supporting cultural and entertainment spaces complement the main function of the center. The cinema club, music club, art exhibition, cultural products, fashion and entertainment, tourism, catering services, theme bars and other support functions provide the visitors with diversified services, and extend the culture of the Expo center cultural extension. The public space and landscape platform make full use of waterfront and surrounding landscape resources.

The World Expo Culture Center will display and continue the theme, "better city, better life", lead the sustainable development of cultural industry, to become an 24-hour operation multi-functional modern culture and entertainment area with distinct characteristics, variability, openness etc. In the overall layout, the floating saucer-shaped main building is situated at the center of the site. The ground single-layer base around the main building serves as commercial space and auxiliary space, contacting with the surrounding landscape. The top platform and grass ramp on the base serves as the evacuation pass for the users in the main building, and have entertainment function. The base and the main building reflect primary and secondary distinct and organic integration relationship. The boiler room and cooling tower are placed on the east side of the building.

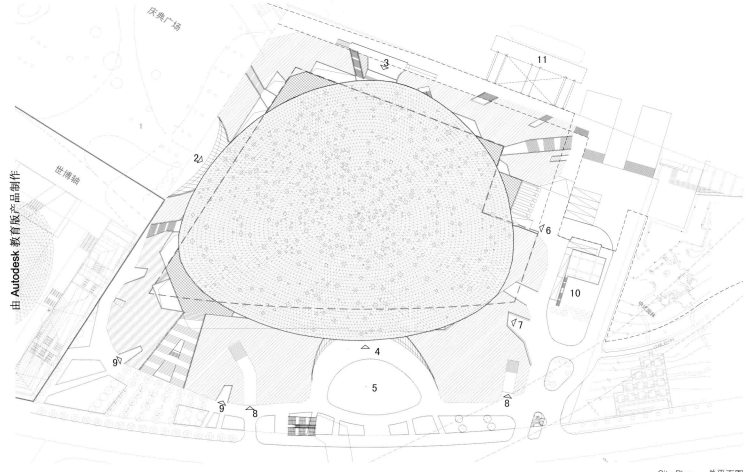

Site Plan   总平面图

　　上海世博文化中心是世博会最重要的永久性场馆之一，在世博会期间承担各类大型演出和活动，满足世博会大型文艺演出需求。同时，还充分考虑世博会会后的后续利用和可持续发展需要。

　　演艺中心主体为 18 000 座的多功能演艺空间，通过灵活分隔等技术手段，可形成不同规模和形态的观演空间，能满足大中型综艺演出、体育赛事、集会庆典等多功能的使用需求。而配套文化休闲娱乐空间构成了对世博文化中心主体功能的补充，电影俱乐部、音乐俱乐部、艺术展示、文化产品销售、时尚娱乐、旅游观赏、餐饮服务、主题酒吧等配套功能为观众提供了多样化的服务，扩展了世博文化中心的文化外延。公共空间及空中观景平台充分利用了滨江和周边的景观资源。

　　世博文化中心将展示和延续"城市，让生活更美好"的主题，引领文化产业的持续发展，实现 24 小时全天候运营，具有鲜明的演艺特征，集可变性、开放性、娱乐性和可经营性为一体的多功能现代文化娱乐集聚区。在总体布局上，将悬浮的"飞碟"状主体建筑置于场地的中部。围绕主体建筑的地面单层基座作为商业空间和辅助空间，以大面积草坡覆盖，与周边场地的景观融为一体。基座顶部平台及草坡用于主体建筑中人员的疏散，并具活动与观景休闲的功能。基座与主体建筑间体现出主次分明而又有机融合的整体关系。主体东侧设置锅炉房、冷却塔。

HONG KONG LINGNAN UNIVERSITY COMMUNITY COLLEGE
岭南大学社区学院

Architects: Wang Weijen Architecture/ AD + RG
Location: Tuen Mun, New Territory, Hong Kong
Site area: 10,800 m²

设计机构：王维仁建筑设计研究室 /AD + RG
项目地点：中国香港特别行政区
总用地面积：10 800 平方米

This project is one of a series of design researches that the architects have been working on: how to re-invent the traditional courtyard typology for contemporary multi-leveled urban conditions. It also aims to establish an innovative fabric system of classrooms to create a variety of open spaces at different levels. In the humid and hot weather of Southern China, it also provides shaded open spaces and opens up several two leveled openings, allowing the cross ventilation to travel across the courtyards.

Rather than using large scale of tall blocks that conventional college design would do, this design identifies two-leveled linear module of classroom blocks, interlocks, rotate and overlap them to create a series of two-storied courtyard unit of 14 by 14 meters. In addition to a sequence of connected courtyards on the ground floor, such layout enables most classrooms on the upper levels to share a sense of sky-courtyard and have a terrace deck near-by each classroom courtyard. For a college that is nearly seven-story-high and located on a sloped site, the design moderates the scale of spaces and establishes identities. By such arrangement the design also lifts most of the building mass from the ground and preserves the original contour and green slope, as well as keeping all original trees on the site.

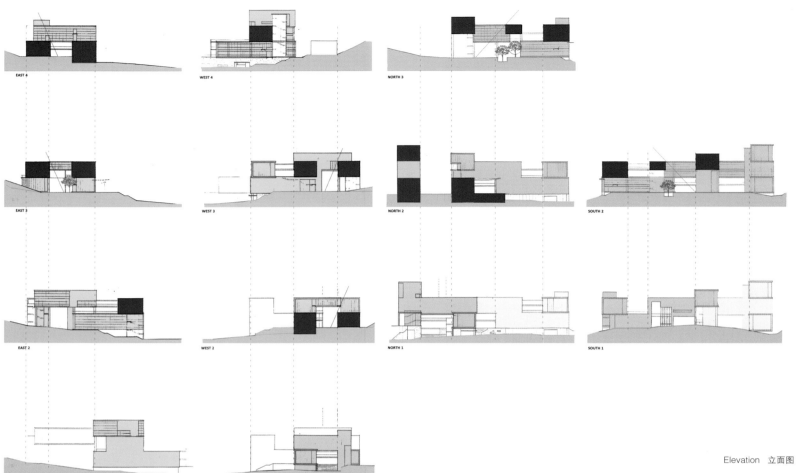

EAST 4 WEST 4 NORTH 3

EAST 3 WEST 3 NORTH 2 SOUTH 2

EAST 2 WEST 2 NORTH 1 SOUTH 1

Elevation 立面图

这个项目是建筑师在现代高密度的多层城市环境中体验中国传统庭院空间关系的系列作品之一。这个设计还旨在建立一个创新的课室肌理系统，在不同的层面创造多样性的开放空间。在气候潮湿炎热的中国南方，这种设计还提供了一些阴凉的开放空间，以及由体量错置后形成的大开口所引进的空气，可以形成对流的穿堂风穿越庭院和室内。

与传统大学使用大尺度的高层建筑体量不同的是，本设计以两层长条形单边走廊的课室体块作为基本模块，将它们相互交错、旋转和叠加，形成了一系列 14 米 × 14 米的两层合院。除了一楼有一系列的庭院外，这样的布局使上层的大多数教室可以共享一个空中花园，并且在每个教室庭院的附近设置一个平台。对于一个有七层高、且坐落在斜坡上的大学来说，这种设计缓和了空间比例，强调了自身的可识别性。这个设计将建筑主要体量托高而脱离地面，以保留原有的草坡地形，使原来的树木还可以在原地生长。

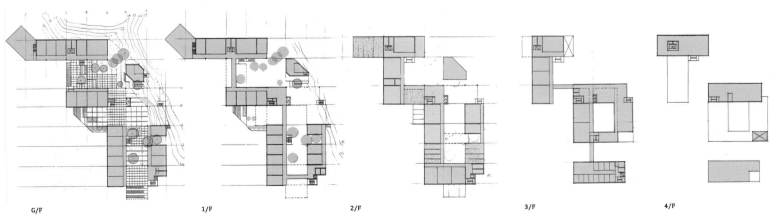

G/F   1/F   2/F   3/F   4/F

Plan　平面图

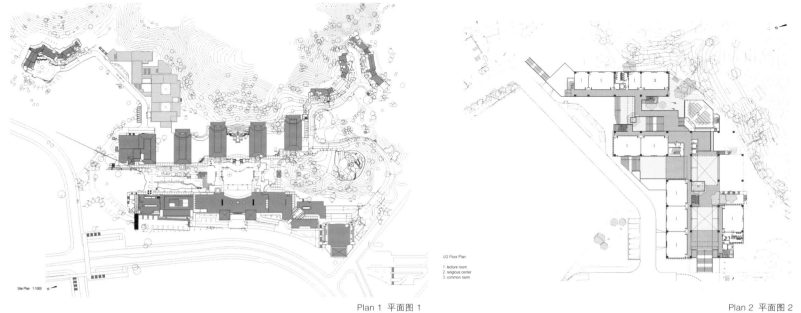

Site Plan 1:1000

UG Floor Plan
1. lecture room
2. religious center
3. common room

Plan 1 平面图 1   Plan 2 平面图 2

Modelling 模型图

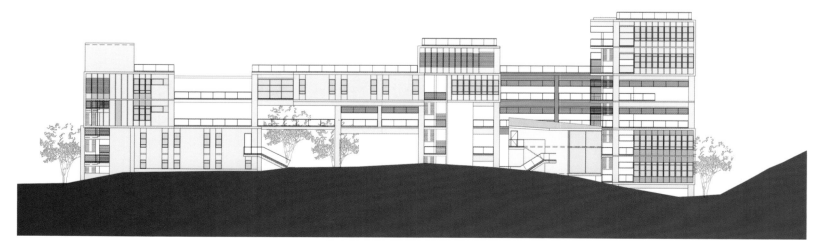

South Elevation 南立面图

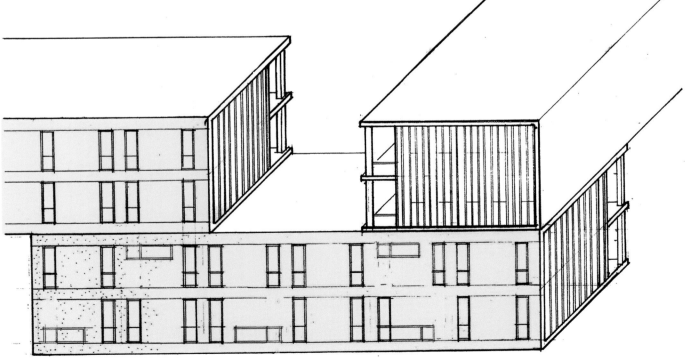

TYP. CLASSROOM ELEV
Lingnan 30/12/02 1/200

Elevation 立面图

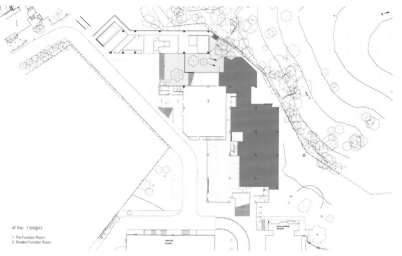

Underground Floor Plan 1　地下平面图1

Underground Floor Plan 2　地下平面图2

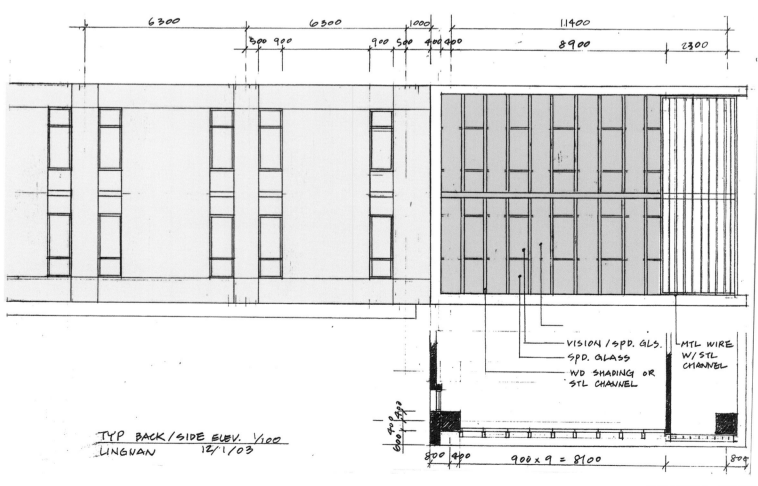

TYP BACK/SIDE ELEV. 1/100
LINGNAN   12/1/03

VISION / SPD. GLS.
SPD. GLASS
WD SHADING OR
STL CHANNEL
MTL WIRE
W/ STL
CHANNEL

Detail Drawing　细节草图

SHANGHAI EXPO CENTER
世博中心

Architects: ECADI
Construction: Shanghai World Expo Group Co.,Ltd.
Location: Shanghai, China
Site area: 66,500 m²
Gross floor area: 141,990 m²

设计机构：华东建筑设计研究院
建设单位：上海世博（集团）有限公司
项目地点：中国上海市
总用地面积：66 500 平方米
总建筑面积：141 990 平方米

As one of the most important permanent buildings for the World Expo, the Shanghai World Expo Center is the core functional place for important foreign affairs and government affairs such as grand ceremony, reception, forum, press conference and large-scale performance etc. After the Expo, it was restructured to be a first-class international conference center. The building has the advanced and complete facilities, graceful and demure architectural form, and colorful interior space. It is the crystallization of technology and art with classic timeless quality and new concept, standing the first position among the green and low-carbon buildings. The whole building adopts the permeable square and vegetative roof, solar photovoltaic technology, program-controlled green space irrigation and wastewater reuse system, pneumatic garbage collection and water circulating cooling technology. Unique semi-conductor white light illuminating and the inflatable hollow metal wire composite glass curtain wall are used to embody the main image of green energy-saving building, and has reached the international leading level of intelligent ecological architecture. The Expo Center has become China's first largest three-star green public buildings, and the first Expo building which has gained the LEED Gold Certification.

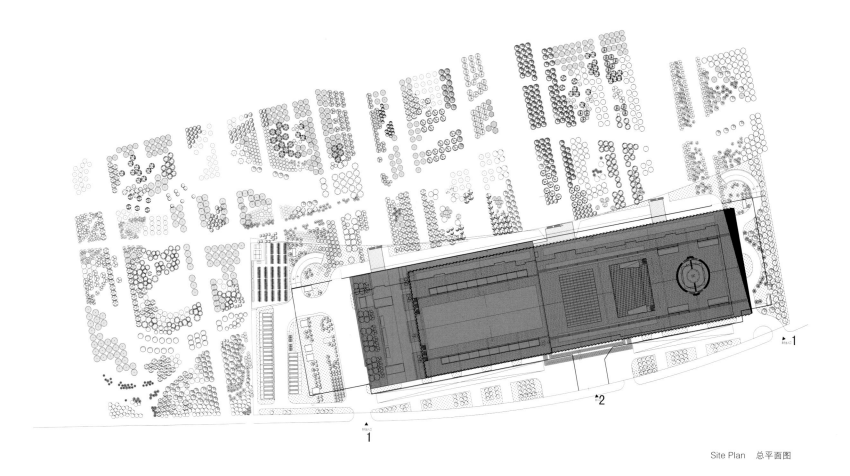

Site Plan　总平面图

作为上海世博会最富影响力的永久建筑之一，世博中心是世博会举行重大仪式、礼宾接待、高峰论坛、新闻发布以及大型演出等重要外事和政务活动的核心功能场所，世博会后则转型成为一流的国际会议中心。其先进完备的功能设施、大气端庄的建筑形态及缤彩纷呈的内部空间，均堪称技术与艺术的结晶，既具备经典永恒的气质，更体现全新的理念，走在绿色低碳建筑的最前列。

整个建筑采用全透水道路广场和屋顶植被、太阳能光伏技术、程控型绿地微灌和中水系统、气动垃圾回收和江水循环降温技术。幕墙运用了独具特色的半导体白光功能照明设备和充气中空金属丝复合玻璃材质，更体现了绿色节能建筑的主要形象和标志，达到国际领先的智能化生态建筑技术集成水平。世博中心已成为中国首座规模最大的绿色三星级公共建筑，也是世博会有史以来第一个获得"美国绿色建筑分级系统"——LEED金奖认证的世博会建筑。

CAPITAL NEW AIRPORT HOTEL BEIJING

北京新国际机场酒店

Architects: WSP Architects
Location: Beijing, China
Site area: 145,275 m²

设计机构：维思平建筑设计
项目地点：中国北京市
总用地面积：145 275 平方米

With a specific location, Beijing Capital Airport Hotel is right at the south side of airport express station of Beijing Capital Airport Terminal 3. The hotel block is symmetrically divided into two parts and the height limit of the buildings at both sides is 35 m. Thanks to this special location, the Capital Airport Hotel becomes an important reception building to visitors to Beijing. From here, people come to know Beijing and China. The design concept is based on the transport carrier function of the airport. It endows the hotel with the image of "Portal of China" and reflects its role of portal to the whole world in the thought of both exposure and enclosure.

The two blocks are for two different hotel operators. However, they integrate the whole site through a common structural grid. The design is based on the street, yard and lane elements in the city life, which are the simplified and transformed quadrangle dwellings and alleyways—the basic elements in Beijing city fabric. These different courtyards complement and echo to each other, forming the building compound. They merged into the prosperous city life and meanwhile become the extension of Beijing city texture.

The design demonstrates the quality of sublimity and principle of energy conservation, modularization does not only express the coordination between human scale and dimension, between overall layout and carefully designed architectural details, but also achieve a grand expression for the new landmark of the city.

The lobby of the airport plaza has glass as its surface, with the intention to reflect the graceful skyline of airport terminal. The main body of the building is finished with light-colored stone cladding. External window of the main building is double glazed wooden window. The skylight system on the roof, gives transparence and openness to the public part of the building. Division on elevation and podium roof is consistent and yet rich in materials. Stones, metals and glasses are carefully composed to create an integral facade with a harmonious sense of grandness and monumentality.

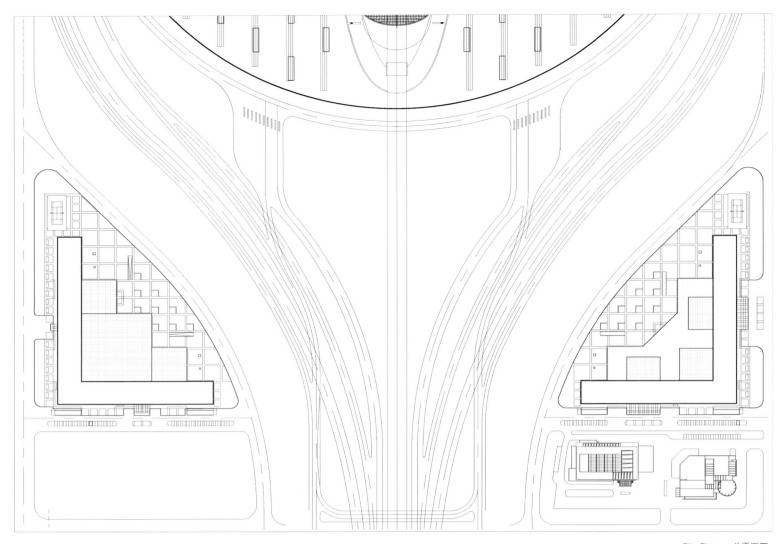

Site Plan　总平面图

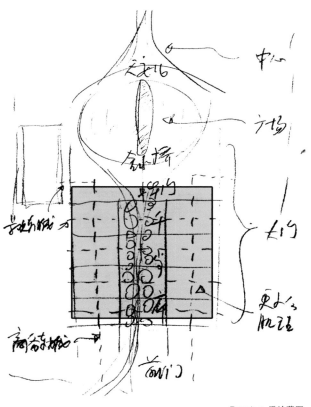

Drawing 手绘草图

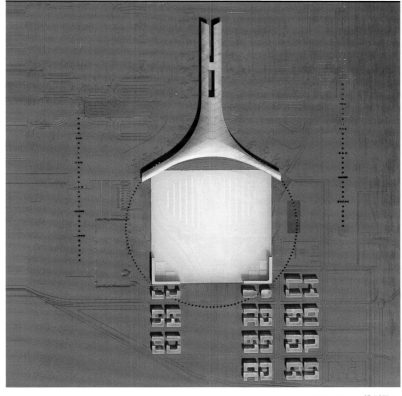

Modelling 1 模型图 1

北京首都机场酒店地理位置独特,位于北京首都机场 3 号航站楼轻轨站南侧,地块以中轴对称为基准被分成两个部分,两侧的规划建筑限高 35 米。凭借这样独特的地理位置,首都机场酒店成为旅客到达北京后首先映入眼帘的一座完整建筑,人们可以通过它来了解北京,甚至是中国。设计概念依据飞机场的交通载体功能,赋予了其"中国门"的形象意义,以开放而又封闭的立意来表现其作为通往世界各地之门户的地位。

建筑用地的两个区块由两组不同的酒店运营商负责,但他们选用一种共同的结构肌理统一整个酒店基地,设计提取了城市生活中普遍的街区院巷元素(这些也是北京城市肌理构成的基本元素——四合院和胡同的变体),将其简化、变形,使这组建筑被不同的庭院有机组合在一起,互补互生,融合成繁荣的城市生活,同时也作为北京城市肌理的新生延续。

设计中以简约、节能、生态和环保作为设计的基本原则,模数化的设计不仅仅体现于人体舒适度和设计尺度的协调统一性,完整和谐的整体格局与精心设计的建筑细节在充分体现建筑趋于理性主义的同时,还充分展现了对人性的全面关怀,赋予整个建筑和谐、富于人情、简约与纯净的立面风格,使建筑成为城市的新地标。

机场广场一侧的大堂部分以玻璃为表皮,意图映射出机场航站楼与高架桥的优美轮廓,建筑主体用浅色干挂石材幕墙,节能环保;主体外窗采用带有双层中空玻璃的德式木窗;屋顶采用玻璃天窗系统,体现了建筑公共区域的通透性与开放性。在造型设计上力求通过风格一致的立面处理使不同功能的几部分达成统一,在立面及裙房屋面上的划分保持一致。通过材料的变化,石材、金属和玻璃的巧妙组合形成丰富的虚实对比变化,并形成统一的整体,充分发挥了材料各自的象征意味,削弱了巨大的体量造成的压抑感,获得了轻盈的视觉效果。

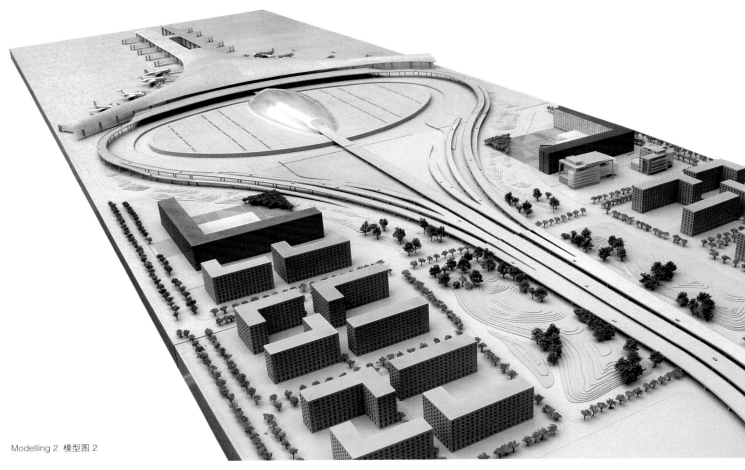

Modelling 2 模型图 2

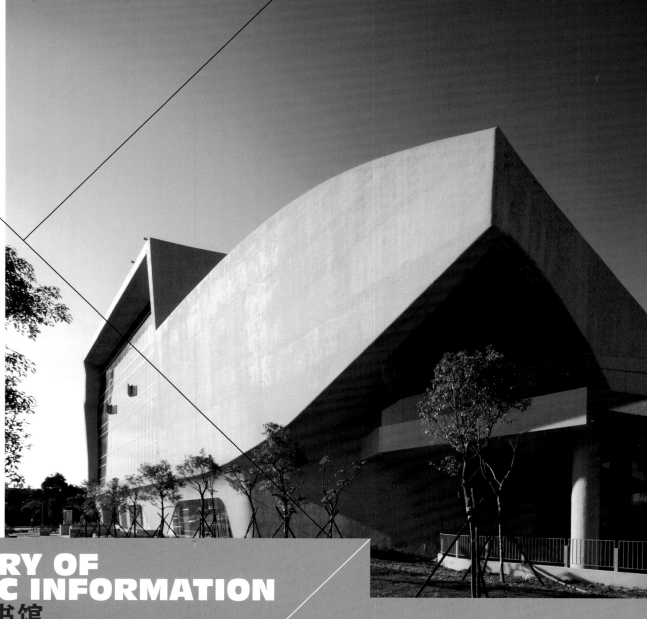

LIBRARY OF PUBLIC INFORMATION
公共资讯图书馆

Architects: J.J. Pan & Partners, Architects & Planners(JJPan)
Location: Taichung, Taiwan, China
Site area : 41,797 m²

设计机构：潘冀联合建筑
项目地点：中国台湾台中市
总用地面积：41 797 平方米

The design concept of the national library of public information originates from the urban texture of the ditch in Taichung city digged in 18th century with "horizontal flow" as the development theme of this cultural landscape architecture; it tries to take flowing facade folds and meandering flowing morphology as a symbol of "knowledge" to remind the public of "searching for a city, reading a city".
In 1920 Corbusier gradually developed "five points of the new building": independent columns, roof garden, free plane, horizontal windows and free facade. This museum is in accordance with the above requirements but don't deliberate to imitate modernism. In addition to achieve the building functions, it makes an attempt on another possibility no matter facade or interior to produce more space interactive. At the same time, it reflects the readers'attitude to ways to use library mainly with electronic information and multimedia services. "L" shaped configuration and the natural arc green belt landscape formed with one side of the green park road concaved show welcome to people.
The facade design of National Public Information Library abandons traditional square straight modeling.Taking natural curved lines, symbolizing the transfer of perpetual creation and accumulation of knowledge;It has eight sizes of mosaic pattern composed external walls, such as riverbed pebbles liked sunlight refraction, which make the architecture emerged with different expressions in each period of the day, and even more dramatic tension in the shape of the mushroom -shaped column of cyclist. Readers enjoy the atrium landscape and green green park road through the flow curved wall and horizontal windows design, which make the architecture no longer parallel with nature,but respect for the natural life and context,meantime make a vigorous response to the base environment.

1   Lobby
2   Digital experience zone
3   Coffee shop
4   Convenient store
5   Bookstore
6   Automated book return
7   Catalogues / Searching
8   Digital children's center
9   Children's reading zone

0   10   20m

First Floor Plan   1层平面图

Reading area introduce natural light by sunroof and side horizontal ribbon windows. The concept of roof dot window reduce artificial lighting that has energy saving effect at the same time.Large number of tree species in the park use Taiwan 's native planting.Using recycled rainwater to set the landscape pool in order to form microclimate, adjusting the ground temperature, and combine with a green roof garden, looking forward to improving the urban heat island effect.After the library built, it is not only the spirit hall that heritage national cultural assets , heritage intellectual activity , and a place that allows people living and learning,but also become one of the important landmarks in Taichung.

设计说明

国立公共资讯图书馆主要设计概念源自于台中市 18 世纪开凿的水圳的都市纹理，以「水平流向」为此一文化地景建筑之发展主题；试图以流动的立面皮褶、蜿蜒流动的形貌，以象征「知识之流」的样貌，提醒市民「寻找一座城市，阅读一座城市」。

柯比意在 1920 年代逐渐发展出「新建筑的五项要点」：独立柱、屋顶花园、自由平面、水平窗带、自由立面。本馆符合以上条件，但并非刻意模仿现代主义，而是企图在实现建筑功能之外，尝试另一种可能性，不管在外观或是内部，让这种可能性产生更多空间互动，同时反映现代读者对待以电子化信息及多媒体服务为主的图书馆的使用方式。"L"形配置、临绿园道一侧内凹形成自然弧线的绿带地景，对人们展现出欢迎之意。

国立公共资讯图书馆外观设计上摒弃传统四方直线的造型，采取弧形自然的线条，象征着流转不息的创作与知识累积；八种尺寸不同的马赛克图案组成外墙，如河床卵石般的对应阳光的折射，让建筑在一天中的每个时段都呈现出不同表情，并在葦状柱的造型烘托下，更显戏剧张力。读者透过曲线流动之墙面及水平窗带设计，享受中庭景观及绿园道的绿意，使建筑与自然不再是互斥的并行线，而是尊重自然的生命与脉络，并对基地环境做出充满活力的回响。

阅读区以天窗及侧面水平带状窗引入自然光，屋顶圆点开窗的概念同时达到减少人工照明的省能效果。全区公园树种大量采用台湾原生植栽。运用回收雨水设置景观水池而形成微气候，调节地面温度，结合屋顶绿化花园，期待能改善都市热岛效应。本图书馆建筑成后，不仅是传承国家级文化资产、知识活动及民众生活学习的精神殿堂，同时也成为台中市的重要地标之一。

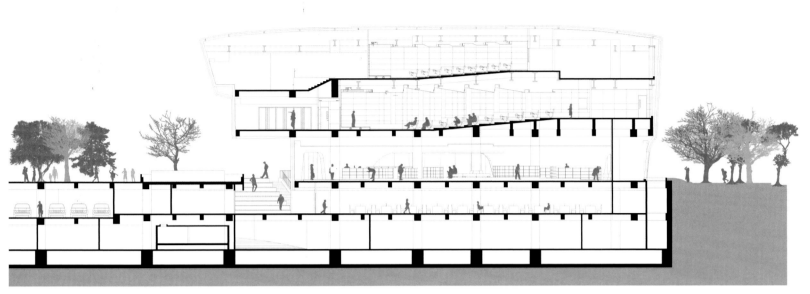

Elevation　立面图

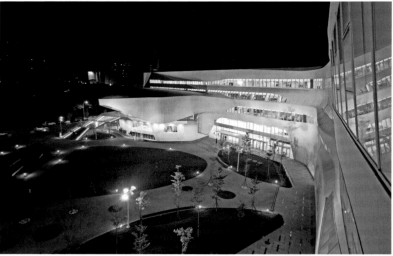

HUNYA CHOCOLATE MUSEUM
宏亚巧克力博物馆

Architects: J.J. Pan & Partners, Architects & Planners
Location: Taoyuan, Taiwan, China
Site area: 3,967 m²

设计机构：潘冀联合建筑
项目地点：中国台湾桃园县
总用地面积：3 967 平方米

The site is surrounded by agricultural land and commands an open view with local residences encircling the peripheral area. The museum sitting on the north-eastern corner of the site and the factory on the south-western side makes rooms for a largest possible outdoor garden. Experience in the museum provides tourists with knowledge of chocolate, and a tour around the factory familiarizes them with the chocolate-making process. The string of activities in the museum, plaza, and factory presents a most fulfilling trip to the world of chocolate.

Hunya is a renowned food brand in Taiwan with a long history. To transform the chocolate brand image and inspire innovations, the building adopts solid volume cut in different angles with the chocolate-colored exterior to convey the imagery of chocolate. Through several major cracks in the appearance a totally distinct, flowing inner space is seen. A void space is created by the 3-floor atrium through the center of the volume, with a sky-bridge and staircases crossing within. The terminal point of tour is a greenhouse of cacao trees. Various exhibitions take place in the spaces on both sides of the greenhouse, like chocolate, a symbol of ever-lasting taste with surprising fillings, embodying the daring, innovative brand spirit.

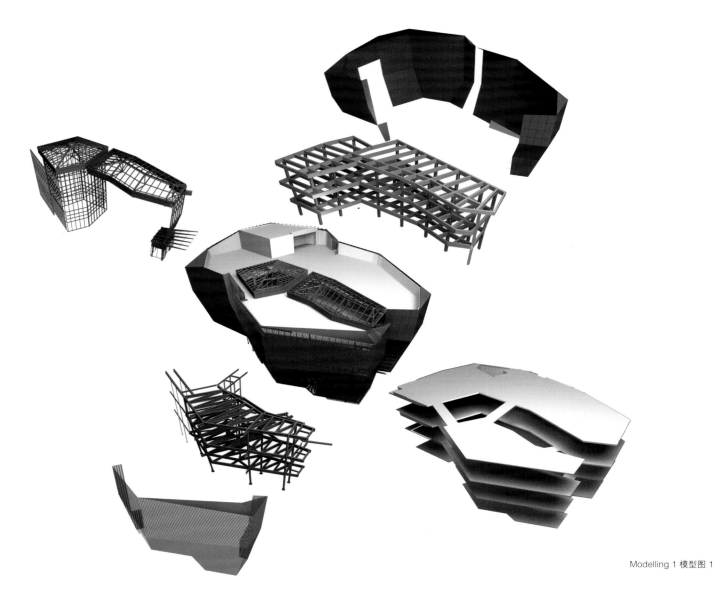

Modelling 1 模型图 1

项目周边为一般农业用地，视野开阔，外围之外则是一圈民房。博物馆配置在基地东北角，与西南方的工厂间留出最大的户外广场范围。博物馆提供游客巧克力的相关知识，巧克力工厂的导览则让游客了解巧克力的体验制作过程。透过参观博物馆、广场、工厂的活动彼此串联，以期给游客最完整的巧克力体验之旅。

　　多年来，宏亚是台湾传统且极具口碑的知名食品品牌。由于巧克力品牌形象的转型以及创新，本案外观设计上采取厚实、被切削的量体感以及巧克力色外墙等手法来传达巧克力意象的概念；而透过外观上几个主要的大型切口则能看到内部截然不同且流动的空间，三层楼大挑空形成的虚空间从中划过建筑物，天桥、楼梯穿插其中，视线的端点则是种植有可可树的透明大温室。形形色色的展览活动在两侧的空间举行，象征隽永的巧克力包覆着令人惊喜的馅料，体现大胆创新的品牌精神。

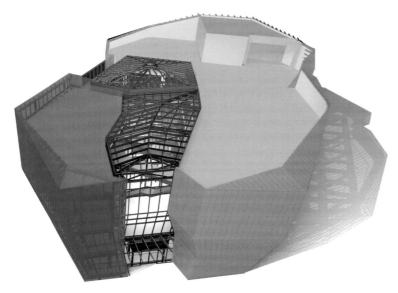

Modelling 2 模型图 2

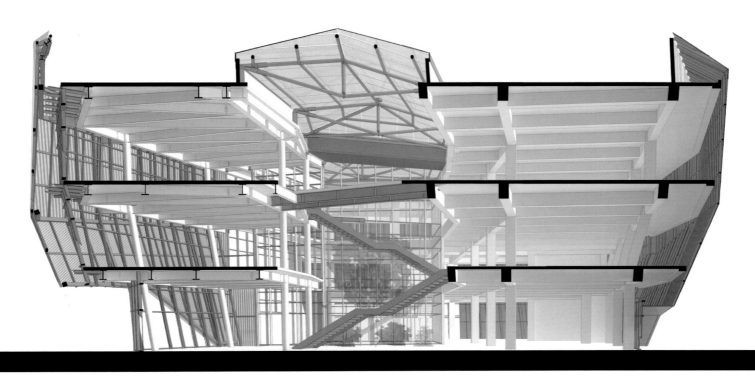

Modelling 3 模型图 3

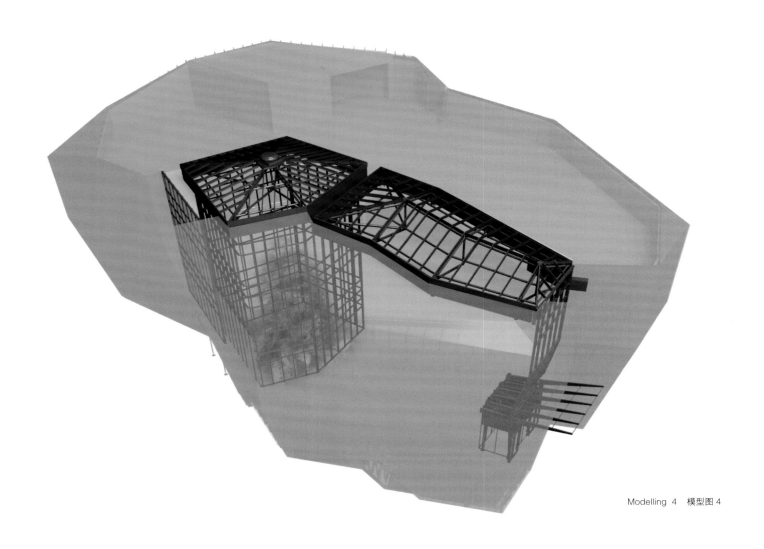

Modelling 4　模型图 4

HUADU DONGFENG STADIUM
花都东风体育馆

Architects: The Architectural Design and Research Institute of Guangdong Province
Location: Guangzhou, China
Site area: 37,000 m²

设计单位：广东省建筑设计研究院
项目地点：中国广州市
总用地面积：37 000 平方米

As a hillside development, the project has its core value lying in the harmony and coexistence with the environment.

We adopt a simple and mellow dew-like ellipsoid shape to respond to the emerald mountains. Thus, the edges and corners are avoided and the conflict with the surrounding architectures, especially the residential buildings is also eliminated, achieving the harmony with the surrounding environment. Meanwhile, the geographical conditions and landscape resources are utilized as much as possible to make it a local landmark with comparatively unique architectural image and spatial characteristics.

On the basis of the ellipsoid shape, the metallic roof panel and glass curtain wall jointly delineate free and dynamic curves to endow the whole building with an elegant and smooth dynamic image.

The integrated design for structure, building skin and interior space strives for the optimal balance between the structure and the architectural form. The material combination of glass, stainless steel plates and fair-faced concrete highlights the simplicity, generosity and elegance of this modern building.

How to innovate the structural system within a limited economic framework has been haunting the engineers and architects for a long time. With reference to the New Barrel Theory, we managed to work out an innovative approach, i.e. Circular Prestress Larg-Span Steel Structure System. This is the first time in China that the circular prestress is applied to the design of a large-span gymnasium.

The gymnasium takes the multi-purpose uses during and after the games into consideration. The flexibility for post-game uses is incorporated into the site design, process design and facilities utilization to facilitate the uses of various functional zones.

The gymnasium maximizes its functions and shape with appropriate technologies and at low costs. Moreover, various effective environmentally-friendly and energy-saving approaches are also employed.

Huadu Gymnasium integrates the geographical features and functions of modern sports buildings, and perfectly combines the architecture form and functionality under the guidance of rationality. It is a successful try to creating a classic rational architecture.

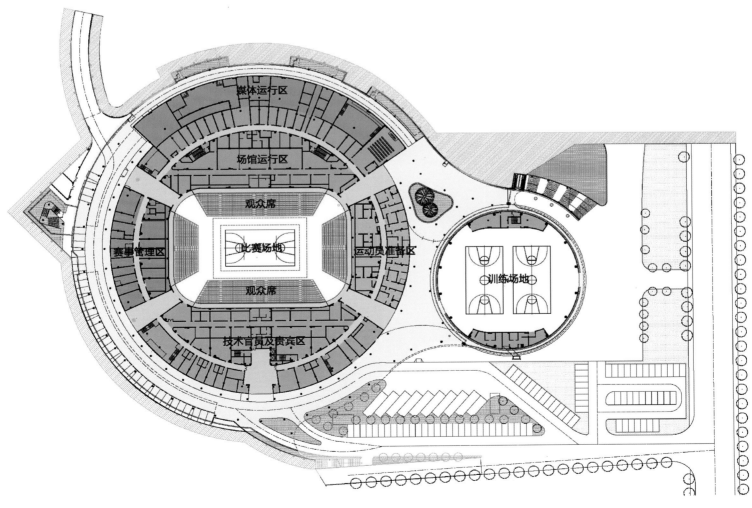

媒体运行区

场馆运行区

观众席

赛事管理区

比赛场地

运动员准备区

训练场地

观众席

技术官员及贵宾区

1st Floor Plan 首层平面图

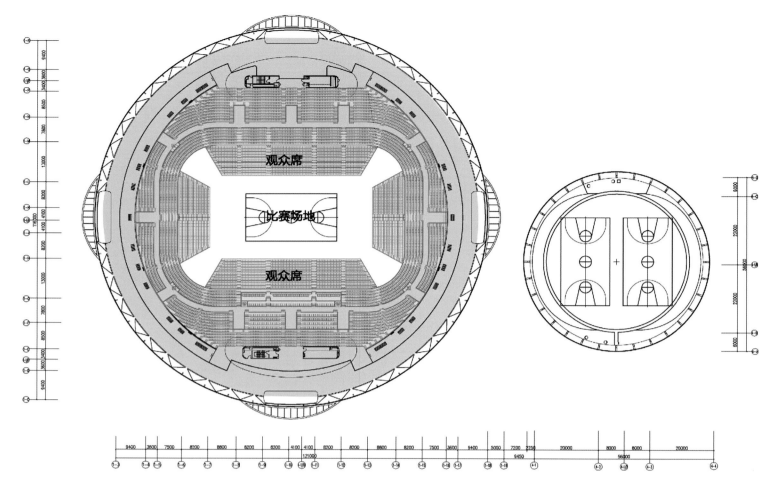

观众席

比赛场地

观众席

3rd Floor Plan　3层平面图

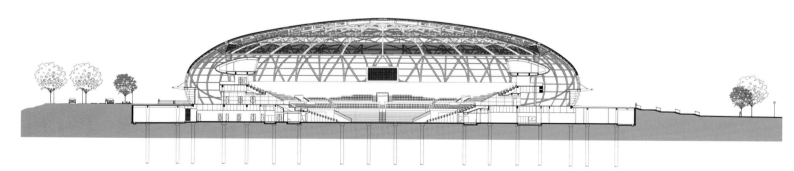

Section 1　剖面图 1

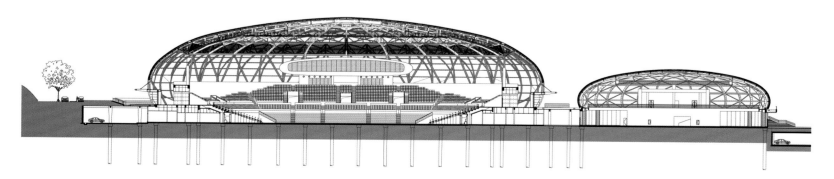

Section 2　剖面图 2

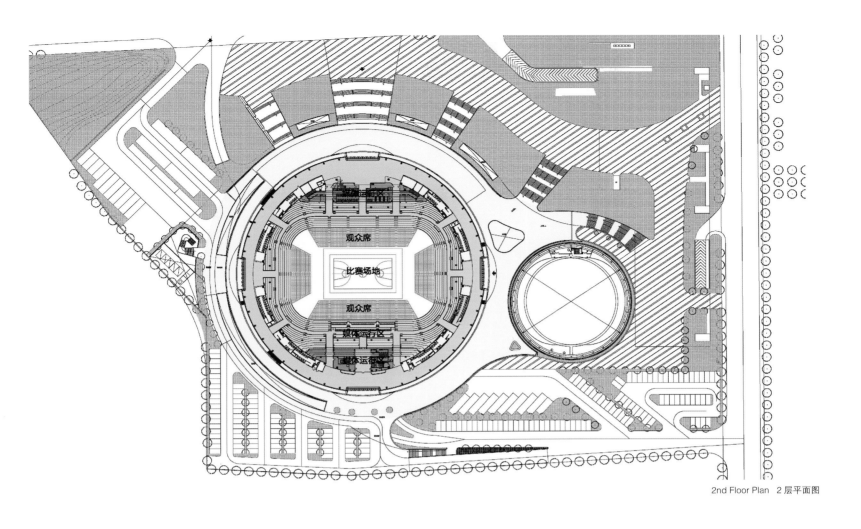

2nd Floor Plan 2 层平面图

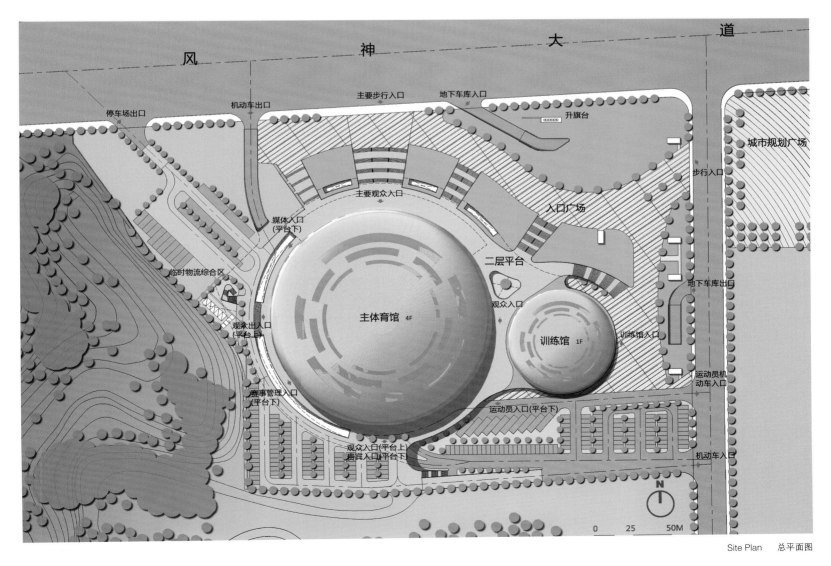

风　　　　神　　　　大　　　　道

停车场出口
机动车出口
主要步行入口
地下车库入口
升旗台
城市规划广场
主要观众入口
步行入口
媒体入口
(平台下)
入口广场
临时物流综合区
二层平台
观众入口
观众出入口
(平台上)
主体育馆 4F
训练馆 1F
训练馆入口
地下车库出口
赛事管理入口
(平台下)
运动员机
动车入口
运动员入口(平台下)
观众入口(平台上)
贵宾入口(平台下)
机动车入口

0   25   50M

Site Plan 总平面图

这是一座位于山边上的建筑，与环境和谐共生是其总体设计的核心价值。

我们将建筑外观设计为一个露珠般圆滑的简单椭球形态，与青山相互映衬。建筑不再有棱角，避免了与周边建筑，特别是居民住宅的矛盾冲突，达到了与周围环境的融洽、和谐。同时也尽可能发挥用地的地域条件及景观资源，以相对独特的建筑形象与空间特点使之成为当地的地标。

在椭圆外形基础上添加了金属屋面板、玻璃幕墙等共同描绘出飞扬动感的曲线，使得整体具有飘逸、空灵、自由流畅的动态形象。

建筑结构与建筑表皮、内部空间的一体化设计，力求结构与建筑形式的最佳平衡。利用玻璃、不锈钢板、清水混凝土等材质组合，凸显建筑的简洁、灵动、大气的现代感。

如何在有限的经济条件下使结构有所创新，是困扰设计者很久的问题。设计师借鉴箍桶原理，创新性地设计出了环行管内预应力大跨度钢结构体系。这是国内首次将环形预应力应用在大跨度场馆设计中。

体育馆设计全面考虑赛时赛后运营的多功能使用。在场地设计、流程设计、设备使用上考虑了赛后利用的灵活性，满足分区使用的需求。

以合宜的技术、低成本最大化实现功能与形式。设计中还采用了大量有效的环保与节能的措施。

花都体育馆是我们在理性精神的指导下，融合地域特征与现代体育建筑功能的需求，实现建筑形式与功能的完美结合，创造理性建筑典范的一次尝试。

CHILDREN'S PALACE AND LIBRARY IN BAOTOU CITY
包头市少年宫、图书馆

Architects: The No. 7 Studio of China Architecture Design & Research Group(Qi studio)
Designer: Cao Xiaoxin
Associate Designer:Zhou Xuan,Shen Xiaolei
Location: Baotou, Inner Mongolia, China
Site area: 52,987 m²
Gross floor area: 52,046 m²
Photographer: Zhang Guangyuan

设计机构：中国建筑设计院器空间工作室
设计师：曹晓昕
合作设计师：周萱、沈晓雷
项目地点：中国内蒙古包头市
总用地面积：52 987 平方米
总建筑面积：52 046 平方米
摄影：张广源

In the project, we focused on the public use of the building and the contribution to the city. The soft boundary between squares and buildings, squares and roads, inner squares and public squares extends the interface. Public space and functional space are permeated by each other, creating a large number of flowing public spaces and multi-use spaces to provide the user with more platforms for various activities.

The design was inspired by the special geology features of continuous grasslands of Inner Mongolia Downs. The site stretches from east to west, forming a 400 meter giant moving curve. The entire building is like growing up from the ground. The curved terrace-backwards design of the library and the children's palace maximally promotes the interaction between the users, public squares and the building.

The atrium extended from inside to outside maximally introduces the sunlight. The open staircase, vestibule, streamlined green areas and pavement section extended to the two building interior enhance spatial visual continuity, and give people a sense of drama vague space. The elevation of the new library, children's palace use a variety of materials: stone, composite ecological plate, aluminum plate, a preembed glass concrete hanging plate etc. All kinds of materials are highly coordinated in a context, to strengthen the integrity and sense of communion. At the same time, they are alone respectively. According to their different functions, the centralized giant spaces of the library and the small units of children's palace combined with the central public square create a different form characteristic. For the children's palace, the internal C-shaped courtyard and central public square form an annular communication space, and in the annular space interface different heights, different colors and different forms of blocks guide the sights to produce all kinds of platform being suitable for children. While the center square as a virtual space can bring out the library volume. Standing at the center of the square, the oncoming thing is stone ramps of different angles on the library west main elevation, which express the strong public sense to the city in a warm and open posture.

在设计中，设计师们着意强调建筑的公共性及对城市的贡献，广场和建筑、广场和道路、内部广场和公共广场之间的柔化界限使相互间的界面不断延续。公共空间和功能空间相互渗透，产生大量流动的公共空间和复合使用空间，从而为使用者的活动提供更多承载的平台。

设计的灵感来自于内蒙古大草原起伏延绵的特殊地貌特征，场地自东向西形成了四百余米巨大的跃动曲线，整个建筑群就像从地面生长出来一样。图书馆和少年宫曲线形的退台设计，最大程度上促进了使用者、公共广场和建筑之间的互动。

由室外延伸至室外的中庭最大限度地导入阳光，开放式的楼梯和连廊以及延伸进两栋建筑内部的流线型绿化区域和铺装部分，增强了空间上的视觉连贯性，为人们提供了一个充满戏剧感的模糊空间。新的图书馆、少年宫外立面运用了多种材料：石材、复合生态板、铝型材、预嵌玻璃砖混凝土挂板等，各种材料特征被高度协调在一个环境下，加强了之间的整体性与交融感。同时它们又彼此独立，根据其各自不同的功能，图书馆为集中式大空间，少年宫多为单元式小空间，结合中央公共广场产生了不同的形式特征。对于少年宫来说，内部"C"形院落与中心公共广场组成了一个环形交流空间，并在环形的空间界面上通过不同高度、不同色彩、不同体块的造型进行视线引导，产生出适合儿童嬉戏游玩的各个平台。而中心广场作为一个虚空间恰好可以衬托出图书馆的体积感，站在中心广场之上迎面而来的是图书馆西侧主立面不同角度的石材缓坡，它们以一种热情开放的姿态向城市传递着强烈的场域公共性。

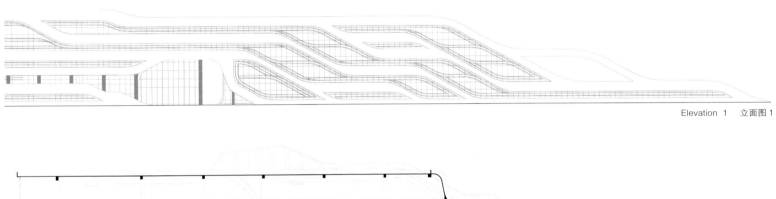

Elevation 1 立面图 1

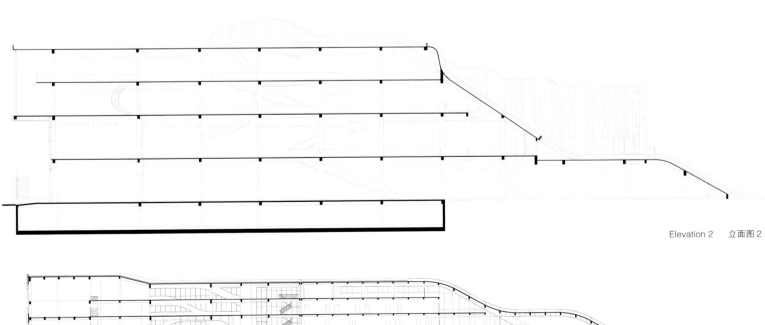

Elevation 2 立面图 2

Elevation 3 立面图 3

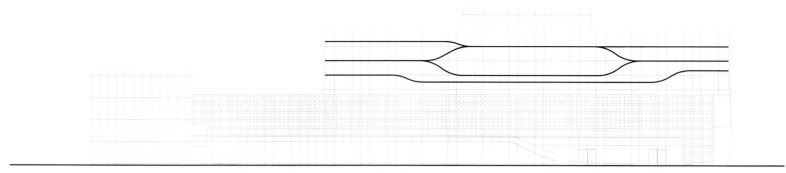

Section 1　剖面图 1

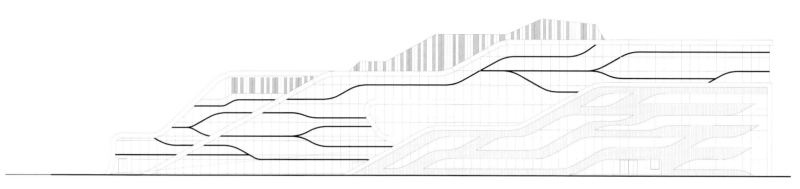

Section 2　剖面图 2

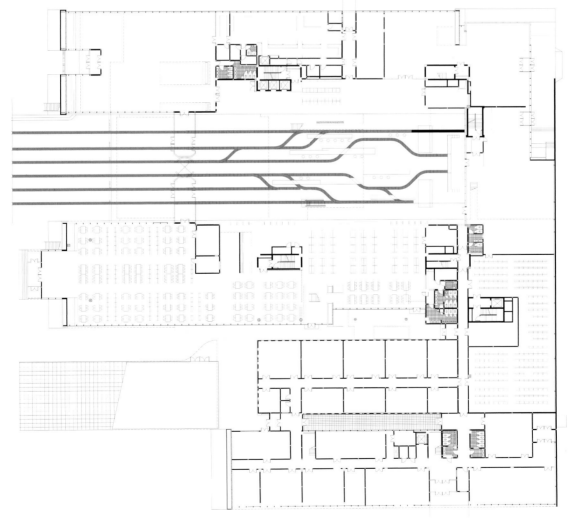

Plan　平面图

XU XIAKE EXPO PARK
徐霞客旅游博览园

Architects: W&R Group
Location: Jiangyin City, Jiangsu Province, China
Site area: 70,000 m²

设计机构：水石国际
项目地点：中国江苏省江阴市
总用地面积：70 000 平方米

Stele forest park
The overall park layout is based on the geographic area through which Xu Xiake has travelled. We integrate the landscape techniques "enjoying broad scenery through a small hole and keeping various scenery with the moving of our steps" in Chinese classical garden, and diversify the calligraphic inscriptions art with a wide range of vertical designs to form unique environment quality.

Museum
The main part of the museum is L-shaped, circling from the southwest corner to the northeast end to form a natural ridge shape. The building achieves a perfect combination with the environment through landscape grass, courtyard, waterscape corridor. The design techniques make full use of the new southern style to produce a museum in which the nature and the artificiality unite, landscape and architecture echo, the tradition and the modern integrate.

Teahouse
Through refining the essence of southern traditional landscape architecture and using the modern and simple techniques of expression, the architecture featuring with new southern style is embodied fully. We use modern techniques and new materials to express the space characteristics and re-interpret the connotation of the traditional culture. Each main functional space's layout is arranged organically and harmoniously on the symbolic landscape form "path" galleries by using opposite scenery and borrowed scenery in landscape architecture design. And the space characteristics of gloriette, pavilions, halls, attics and gallery are expressed. Water courtyard, stone courtyard and bamboo garden separate our sight and bring those landscapes together, whilst integrating the building into the environment.

Master plan 规划图

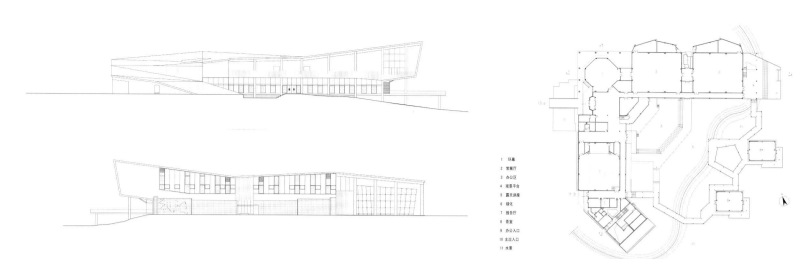

1st Floor Plan 1层平面图

Elevation 立面图

碑林公园
以徐霞客游历的地理区域为线索进行了总体园林布局，将中国古典造园的"小中见大"、"步移景异"等手法运用其间，通过丰富的竖向设计，将书法碑刻艺术予以了多样化体现，形成了极其个性化的环境品质。

博物馆
博物馆主体建筑成"L"形布置，自西南角向东北端盘旋而起，形成自然的山脊造型。
建筑通过景观草坡、景观庭院、水景走廊，与环境完美融合。设计手法采用新江南风格，塑造出将自然与人工相结合、景观与建筑相呼应、传统与现代相融合的博物馆建筑。

茶室
通过提炼江南传统园林建筑的精髓，并运用现代简约的表现手法，体现了新江南风格的建筑特色。以现代手法和新材料展现中国江南园林的空间特色，重新诠释传统文化的涵。各个主要功能空间布置采用了园林建筑中对景与借景的设计手法，使之在象征园林形态中的"路径"廊上有机却自然和谐地排布。展现园林"亭"、"台"、"轩"、"榭"、"廊"的空间性格。转角的水院、石院、竹园分隔了视线带来景观的共享，也把建筑融入了环境之中。

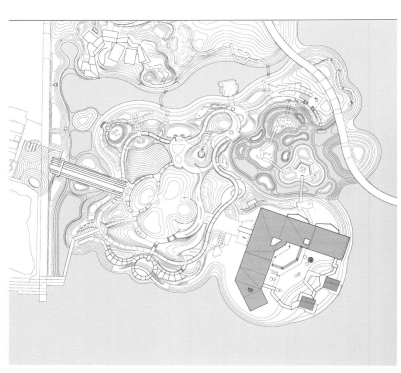

Plan 平面图

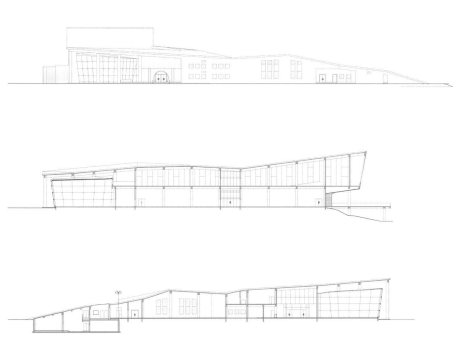

Section　剖面图

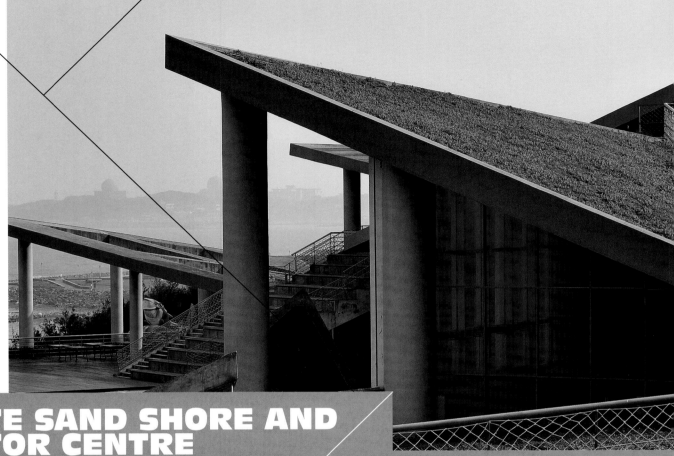

WHITE SAND SHORE AND VISITOR CENTRE

白沙湾海水浴场旅客服务中心

Architects: Wang Weijen Architecture + Beiyu Space Design Studio
Location: Bai Sha Wan, Taipei country, Taiwan, China
Site area: 5,655.82m²

设计机构：王维仁建筑设计研究室 + 北域空间设计事务所
项目地点：中国台湾台北县白沙湾
总用地面积：5 656.82 平方米

Design Vision
The architecture, coordinated with the terrain and the landscape, attempts to change the traditional relationship between person, landscape and the architecture. The architecture is hoped to be a part of the landscape just like the tress and the rock.

Background
The design of this project is supposed to improve the environment of Bai Sha Wan Beach and expand the service facility of the visitor centre by transforming the old buildings of Bei Guan Chu to make it become an important joint of North Coast Travel Series.

Design Ideas

1.Architecture Transformation and Landscape Construction

The first strategy of the design is to make use of the new buildings to connect the old buildings and the surrounding terrain to make the original buildings become a part of the terrain. At the same time the new buildings' folded plate can be regarded as the western side and northeastern side corridor of the old buildings to provide the function of shade and rain canopy.

2.Activity Order and Corridor Square

The second strategy of the design is planning the function order conform to the falling terrain from the road to the coast along the contour. Besides, the design is centered on the passengers.

The design also considers the benign interaction between the passenger center and the adjacent villages to arrange the reasonable function of tourism bureau Bei Guan Chu office space.

3. Landscape Construction and Architecture Landscape

The third strategy which is also the main design attitude is to create buildings which design the landscape. The roofs of the buildings are composed of triangle and parallelogram to match the terrain, control the inclination and adjust the angel of light easily.

The structure of the roofs of the new buildings is upstand beam construction to display the flat material of fair-face concrete of the indoor ceiling on one hand; on the other hand it can offer enough soil depth for the grass on the roofs.

What is hoped by the designers is that when the passengers can not only enjoy the convenient and carefree coast activities, but also feel the wise and noble architecture landscape and the mood of the sea melting into the sky.

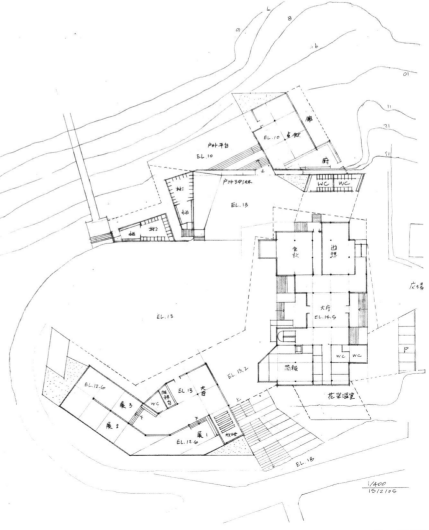

设计构想

建筑配合地形与地景，企图改变人与地景及建筑的传统二元关系，期望建筑像大树、岩石一样成为风景的组合成分。

规划背景

设计希望透过对北观处旧建筑的改造，改善白沙湾海水浴场的周边环境，扩充旅客中心的服务设施，使白沙湾成为台湾北海岸旅游系列的重要节点。

建筑设计构想

1. 建筑转化与地形建构

设计的第一个策略是利用新加的建筑连接旧建筑与周边地形的高程，将原本独立于地景上的既有建筑转变成地形的一部分；设计同时利用新建筑的折板造型，成为旧建筑西侧与东北侧的廊道，提供建筑遮阳与遮雨的功能。

2. 活动序列与廊道广场

设计的第二个策略是顺应着由路面到海岸一路下降的地形，沿着等高线安排序列的相关功能。另一方面，设计以旅客中心为核心。

设计同时考虑了旅客中心与邻近聚落的良性互动，以及安排观光局北观处办公空间的合理功能。

3. 地景建筑与建筑地景

设计的第三个策略，也是最主要的设计态度：希望这是一栋设计地景的建筑。建筑屋面采用三角形和平行四边形的组合，一方面容易配合地形，另一方面易于控制屋面的斜度与调整光线的角度。

新加的建筑屋顶结构采用反梁构造，一方面保持展示室内空间屋面平整的清水混凝土材质，另一方面提供室外屋顶足够的植草覆土深度。

设计者期望的是，游客除了使用便利、淋漓畅快的海岸活动外，也能感受到一些仁者智者建筑景观，海天地一色的心境。

Plan 平面图

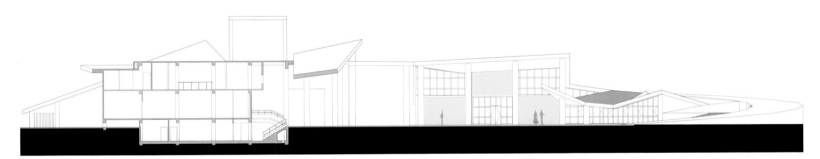

Section　剖面图

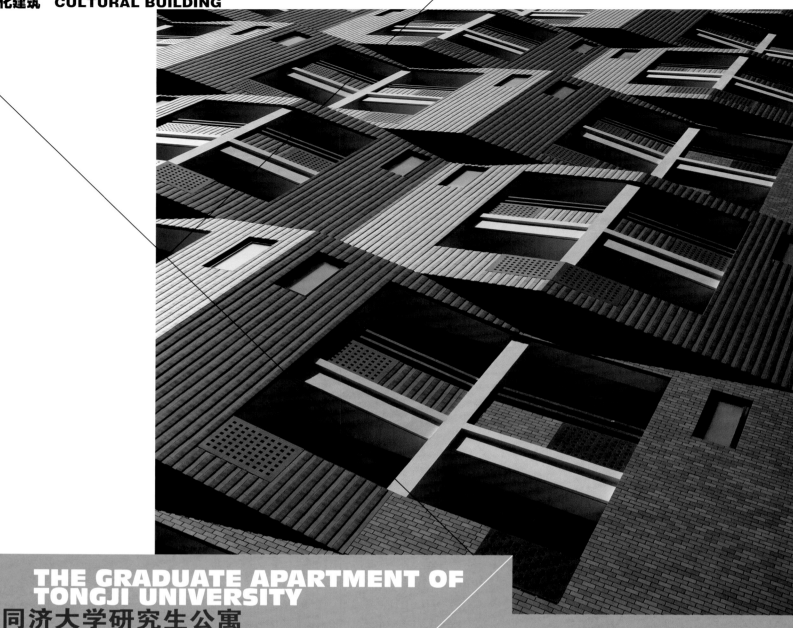

THE GRADUATE APARTMENT OF TONGJI UNIVERSITY
同济大学研究生公寓

Architects: TJAD
Location: Shanghai, China.
Site area: 71,300 m²

建设机构：同济大学建筑设计研究院（集团）有限公司
项目地点：中国上海市
总用地面积：71 300 平方米

The project is located at the Tongji University (Zhaowu District branch school), on its east and south are Yangshupu Harbor and Songhe Park. The building seated in the west campus was been built in 1950s.

The geographical distribution fits the terrain and the texture of the city. It also deals well with the campus space convergence, reduces the feeling of oppression from the High-rise building built in campus. The building is along the river side, interacted with the residential across the river which made the Yangshupu Harbor Space more interesting. The inside of the community avoids the sight barrier between each tower and determines a reasonable construction density. The design also successfully controls the surrounding sunlight effective created by the tower, including the effect on campus plot of east Yangshupu Harbor, which will convenient for the reconstruction. All these should satisfy the requirements of the total number of people living. The difficulty of the design is to balance such many factors. In addition, the design integrated the existing traffic organization in campus, make the south side road of south tower as the car dealer roads, meantime clean the roadside obstacles and strengthen its transport function.

Graphic design uses middle corridor type, the public activity room and the lounge are situated at the end of corridor. All spaces of interior use natural lighting and ventilation. The standard room is for two people and the per capita living area is 8.1 m².

Facade design selects two layers and two rooms as a unit. Combined with these regularly units, the facade gives people a deep impression. Grey is the fundamental color of the architecture, which obtained restrained and rich results.

Designer has done a lot of work in selecting and using materials, in order to strengthen the construction of expression. All tiles are customization and the paste method draws by the designer. After the architects looking and amending the sample at the construction site for many times, negotiating with the material suppliers and construction team, the project has been successfully accomplished with the support of the owners' and everyone's efforts.

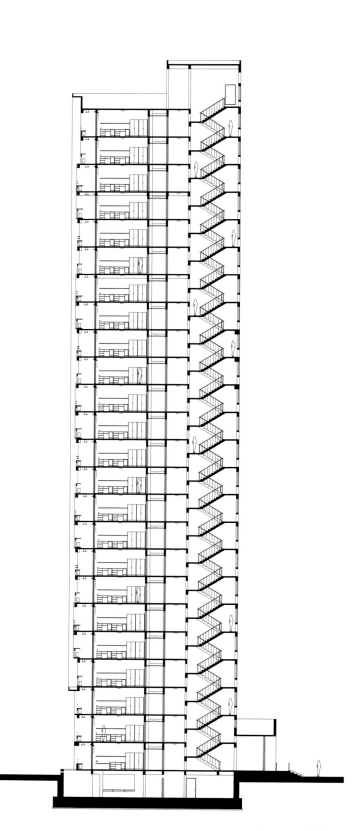

Section 剖面图

本工程位于同济大学彰武路校区，东面是杨树浦港，南面是松鹤公园，西面校园内的教学用房建于20世纪50年代。

建筑布局契合地形，也契合城市的肌理，处理好与校园空间的衔接，减少了高层建筑对校园的压迫感。建筑在沿河的一面，与对岸的住宅楼交相呼应，使杨树浦港两岸的空间更为有趣。社区内部避免了公寓各楼视线的对峙，确定了合理的建筑密度。设计还成功地控制塔楼对周边的日照影响，包括对杨树浦港东面校园地块的影响，为其改造留有较大的余地。而这些都必须满足总居住人数的要求，设计的难度也在于平衡诸多因素。此外还整合了现有校园的交通组织，将现南楼南侧道路作为车行道路，清除路边障碍，强化其交通功能。

平面设计采用内廊式，公共活动室和生活间布置在走廊的端头。内部均自然采光通风，标准间为两人间，人均居室使用面积8.1平方米。

立面设计选取两层两间为一个单元，由这些单元有规律组合而成的外观给人以深刻的印象。建筑色彩以灰色为基调，取得内敛而丰富的效果。

在材料的选择和运用方面作了大量的工作，增强建筑的表现力。所有面砖均为定制，贴法由建筑师绘出。建筑师在工地现场无数次看样、修改，与材料商和施工队磋商，在业主的支持和大家的努力下，得以最终完成。

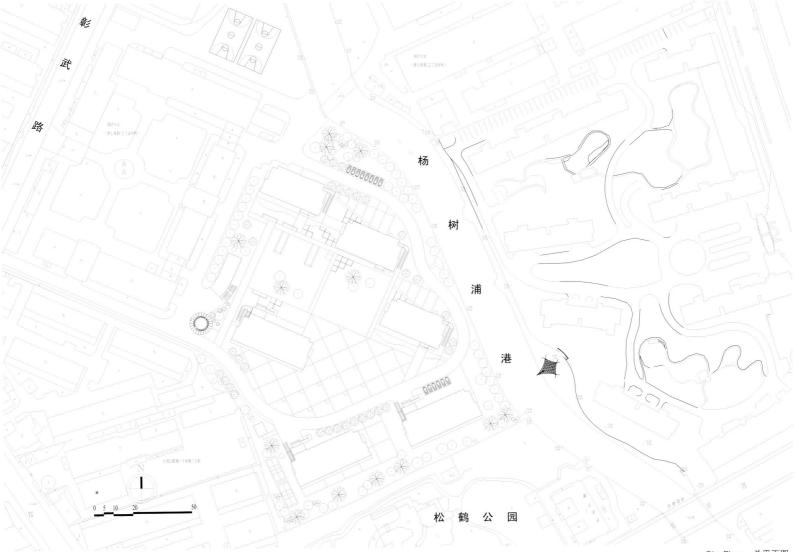

Site Plan　总平面图

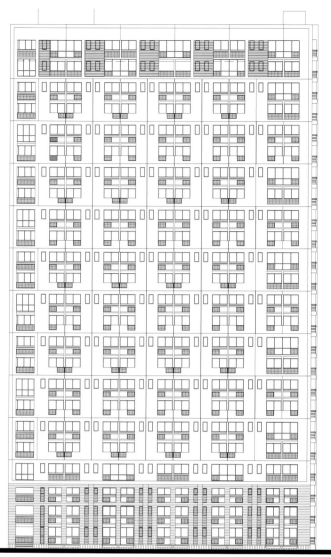

South Elevation　南立面图

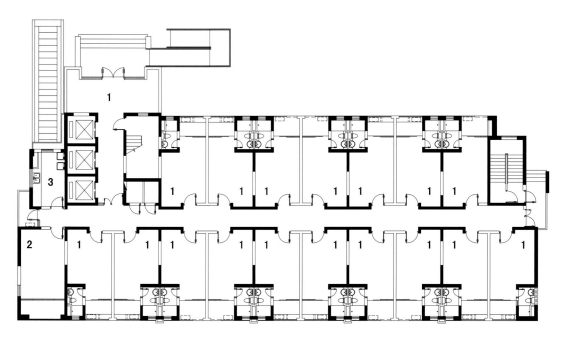

1st Floor Plan   1 层平面图

SHANDONG PROVINCIAL HOSPITAL EASTERN DISTRICT
山东省立医院东院区

Architects: ECADI
Location: Jinan, China
Site area: 95,500m²
Gross floor area: 168,799.5m²

设计机构：华东建筑设计研究院
项目地点：中国济南市
总用地面积：95 500 平方米
总建筑面积：168 799.5 平方米

Shandong Provincial Hospital is a large first class of grade Ⅲ general hospital in Shandong Province, which has a history of hundreds of years. The new eastern hospital district with over 1,500 beds and a daily outpatient capacity of 6,000 people is located at the Olympic Sports Administrative Center in the east of Jinan. Our design pays a close attention to urban planning, medical process, humanistic care and regional culture, which reflects the characteristics of large modern hospitals from 4 perspective as follows: first of all, the adoption of "one axis multi cores" linear planning structure not only integrates the hospital planning into urban planning framework but also ensures the health operation of the hospital planning.

Secondly, clinical specialty as the core of function organization of this district, we build a center with diagnosis, examine, treatment which highlights the technology advantage of the large general hospital, improves the operational efficiency of services and keeps pace with the reform and development trend of the large general hospital.

Thirdly, our design uses natural ventilation, green roof, electric sunshade, automatical communication, office automation, building automation, pneumatic logistics and other environmental friendly, intelligent, and medical technology to raise the management level of modern hospital services.

Fourthly, the shape takes function as the core and the arc podiums hold up four high-rise ward. The ward gable is embedded in the delicate glass block, which achieves a perfect combination of hardness and softness, virtuality and reality, technology and art perfectly, implying the regional cultural characteristics of "full pond of lotus flower, wide spread of hospital spirits".

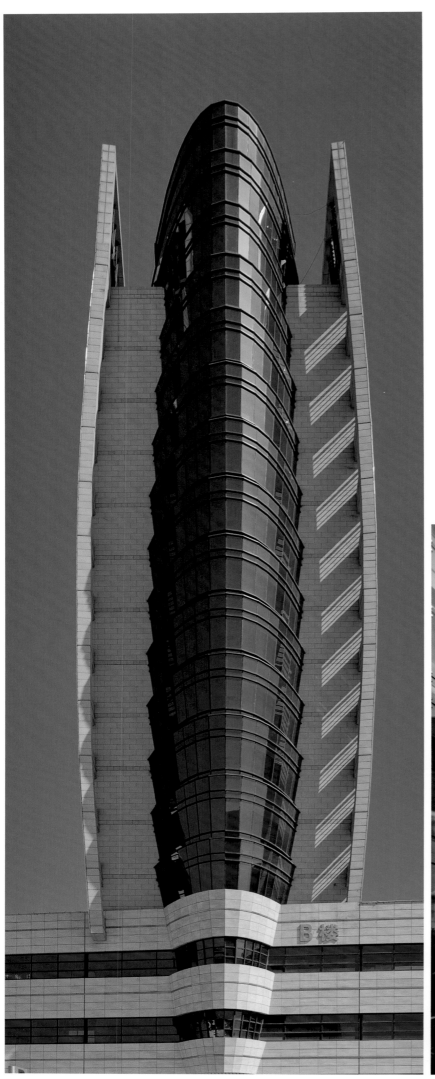

山东省立医院是山东省具有百年历史的大型三级甲等综合医院，新建东院区位于济南东部奥体政务中心，床位1 500个，日门诊量6 000人。

设计关注城市规划、医疗流程、人文关怀和地域文化，从以下4个角度体现大型现代化医院特色：首先，采用"一轴多核"的线形规划结构，使医院规划既融入城市规划框架，又确保医院规划自身的健康运作。

其次，以临床学科为分区功能组织的核心，构建诊、查、治对应的诊疗分中心，突出大型综合医院学科技术优势，提高服务运营效率，契合大型综合医院的改革发展趋势。

第三，设计采用自然通风、屋顶绿化、电动遮阳、通信自动化、办公自动化、楼宇自动化、气动物流等绿色、智能和医疗技术，提升现代化医院服务管理水平。

第四，造型设计以功能为核心，弧形裙房托起四栋高层病房，病房山墙嵌入精致玻璃体块，刚柔并济，虚实结合，实现技术与艺术的完美结合，寓意"荷花满塘，医海绽放"的地域文化特色。

YINCHUAN CULTURE AND ART CENTER
银川文化艺术中心

Architects: China United Zhujian Architecture Design Co.Ltd.
Location: Yinchuan, China
Site area: 33,214 m²

设计机构：中联筑境建筑设计有限公司
项目地点：中国银川市
总用地面积：33 214 平方米

Yinchuan Culture and Art Center is located on the east of Yinchuan City central area, People's Square, faces Grand Theatre on the south, the Guangchang East Road on the west across the library and Shanghai Road on the north across the site. The east side is 30-meter wide road. The project has convenient traffic.

The project design started from October 2005 and completed in October 2008.The site area is 32,943 m², the gross floor area is 33,214 m², FAR is 1.01, the building density is 33.3%, landscaping ratio is 23.4%, the capacity of car parking is 106 cars, and total investment is about 180 million RMB. The whole center includes the Yinchuan Art Museum, Cultural Popularity and Education Center, Art Creation and Study Studio, Yinchuan Culture and Art Academy, art collection room, venue management spaces, cultural supermarket, public rest services, underground facilities etc.

Today, the designs not only pursue the rationality of the traditional function, but also seek the comprehensive driver to achieve the outstanding building. The center is conceived as a compositive entirety, respecting the master plan of the East Area of the square, corresponding with the library and museum. It provides a space for residents to appreciate art and art activities. The building itself is a modern and beautiful artwork, in which citizens can ramble in the art ocean, thus improve the artistic talent of the citizens.

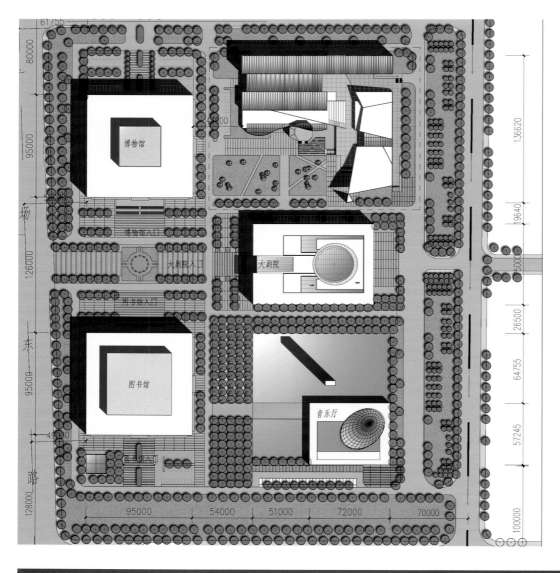

Master Plan 规划图

银川文化艺术中心地处银川城市核心区人民广场东侧，南邻大剧院，西侧与广场东路相邻，其东侧有 30 米宽的规划道路，北面隔规划用地邻上海路，交通便捷。

项目设计自 2005 年 10 月份至 2008 年 10 月份结束。规划设计用地面积 32 943 平方米，总建筑面积 33 214 平方米，容积率为 1.01，建筑密度为 33.3%，绿地率为 23.4%，汽车停车泊位约 106 个，总投资约 1.8 亿元。整个中心包括银川美术馆、群众文化普及教育中心、银川文化艺术院、艺术创研工作室、艺术品收藏室、场馆管理公用场所、文化超市、公共休息服务、地下设备设施等辅助用房。

当今，艺术中心的设计已经不再只是寻求合理性的传统功能，优秀的艺术中心的建筑设计远远超越了这单一的设计动力，因此我们的设计手法是将艺术中心视为一个"综合的整体"，在总体布局上尊重广场东区的整体规划，在尺度、空间、形态上与图书馆、博物馆相呼应。它给市民提供了一个欣赏艺术与艺术活动的场所，并力求使艺术中心本身成为一个造型现代、优美的艺术品，让市民们能在艺术的海洋里徜徉，从而提升全民艺术素质。

SHANGHAI WORLD EXPO THEME PAVILIONS
中国 2010 年上海世博会主题馆

Architects: TJAD
Location: Shanghai, China
Site area: 152,318 m²

设计机构：同济大学建筑设计研究院（集团）有限公司
项目地点：中国上海市
总用地面积：152 318 平方米

Idea Source One: City Neighborhood

Fragment of Neighborhood and City Memory: The main feature of Shanghai texture is its rhythmical fragments of the neighborhood, which created a historic example of happy city life of Shanghai. The themed pavilion roof, having a gross area of 60,000 m², would present the most prominent texture feature in the heart area of World Expo. Combined with City Neighborhood, a large scale repetitive characteristic of the roof stimulates the abstract concept of "Neighborhood Fragments" to be refined to a roofing model during the designing of themed pavilion: By combining hipped plate roofing and solar panel, the large scale roofing forms an effect of criss-cross city and undulating texture, which continues the visional glamour of traditional city space of neighborhood roofing texture and the memory of Shanghai.

Idea Source Two: Ancient Architectural Cornice

Far-reaching Cornice and Traditional Continuity: the main characteristic of traditional Chinese wooden structure is the gray space, possessing both structural and functional features formed under a far-reaching cornice of sloping roof. The structure design of the themed pavilion borrows the idea of "Far-reaching Cornice" from ancient Chinese architecture, which sets main entrance and waiting zone both in its north and south, accordingly projecting large cornices both in north and south. This design abstractly expresses our ancient Chinese roofing feature of "Dougong", or bracket system as prominent subulate steel construction member. The load-carrying members and the top day-lighting panels are respectively settled to replace the stack-up bracket system in ancient architecture. The unique herringbone colonnade forms a contrast to ancient Chinese colonnade through modern architectural language.

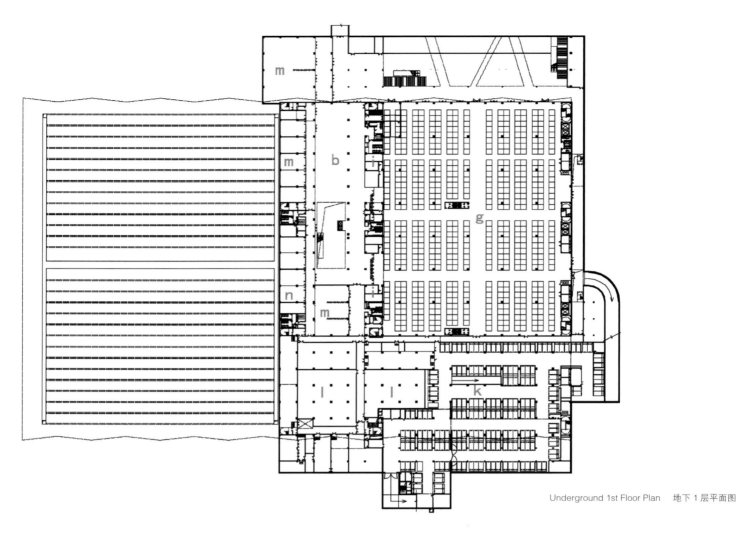

Underground 1st Floor Plan　地下 1 层平面图

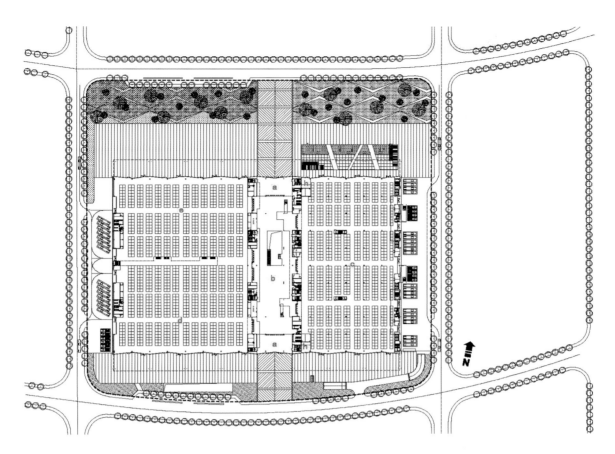

1st Floor Plan 1层平面图

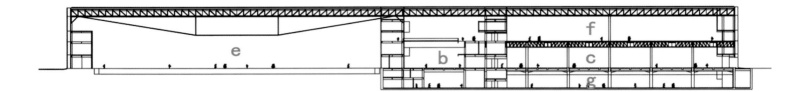

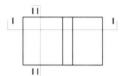

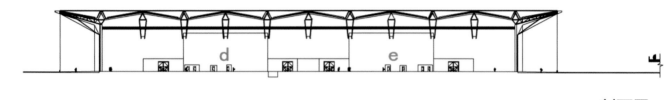

Section 剖面图

理念源泉一：城市里弄

里弄片断与城市记忆：富有韵律感的里弄片断是上海城市肌理的主要特征，是创造上海城市美好生活的历史范例。总面积达6万平方米的主题馆屋面，在世博核心区中将呈现出最为突出的肌理特征。它与城市里弄的大尺度重复性形态特征关联度激发了主题馆设计中将"里弄片断"这种城市意向抽象提炼到屋面造型：利用折板式屋面与太阳能板的组合，将超大尺度的屋面形成了类似城市纵横交错，凹凸起伏的肌理效果，延续了里弄屋面肌理传统城市空间的视觉魅力，承载了上海的城市记忆。

理念源泉二：古建挑檐

出檐深远与传统延续：中国传统木构建筑主要的特征就是出檐深远的坡屋面挑檐下形成的兼具形态、功能特征的灰空间。主题馆形体设计借鉴中国古建"出檐深远"的特点，在南北方向均设主要出入口和等待区，相应地在南北设计大挑檐，将传统古建中檐口特征斗拱抽象演绎成为锥形的钢结构出挑构件，穿孔金属板后依次布置的承重结构构件和顶面的采光板体系，代替了古建中层叠的斗拱形式，造型独特的人字形柱廊通过现代建筑语言与中国古建中的柱廊形成对话。

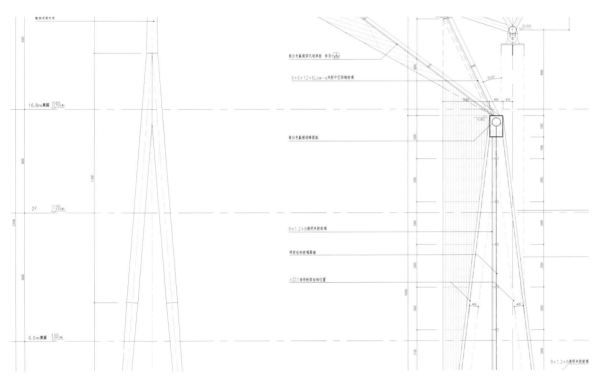

Elevation　立面图

GUANGZHOU ASIAN GAMES HALL
广州亚运馆

Architects: The Architectural Design and Research Institute of Guangdong Province
Location: Guangzhou, China
Site area: 101, 086 m²
Gross Floor Area: 65, 315 m²

设计机构：广东省建筑设计研究院
项目地点：中国广州市
总用地面积：101 086 平方米
总建筑面积：65 p315 平方米

1. Background
The project is located in the southern part of Guangzhou Asian Games Town, being adjacent to the picturesque Lotus Bay. It is a sports venue complex, consisting of a Gymnastic Hall, an Indoor Stadium and an Exhibition Center of the history of Asian Games, totaling a site area of 101,086 m² and a GFA of 65,315 m².
After intense competition, the original conceptual design of A Flowing Ribbon proposed by the Architectural Design & Research Institute of Guangdong Province (GDADRI) won the competition and was implemented.

2. Conception, Form and Space
(1) Design conception: a flowing ribbon—a symbol for the most artistically appealing sport—rhythmic gymnastics
The creative concept is inspired by the game hosted by the Guangzhou Asian Games Town Gymnasium—rhythmic gymnastics. The futuristic and avant-garde architectural image brought out by the creative design of Gymnasium establishes the main venue of the Asian Games as a unique landmark.
(2) Exploration of a new architectural language and experience
The Gymnasium presents a brand new architectural experience which brings the dramatic effect of varying sceneries with changed view-points.
(3) Creation of a proactive, open and comfortable urban public space
The design provides the continuous and flowing non-homogeneous semi-indoor grey spaces and outdoor public spaces in diverse forms, varied hierarchies, thus maximize the publicness of the Gymnasium.

Master plan 规划图

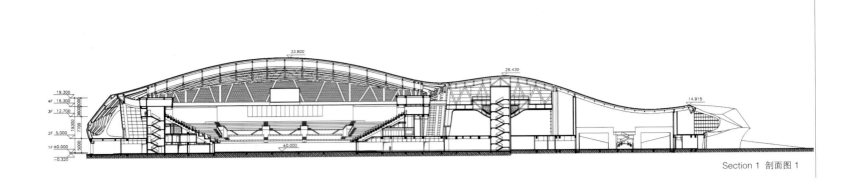

Section 1 剖面图 1

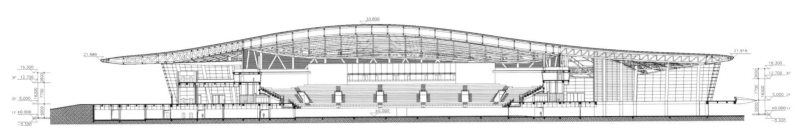

Section 2 剖面图 2

3. Functions and Circulation
The main functional areas of competition venues in the Gymnasium include the audience area, the competition area and the internal functional area. In circulation design, vertical distribution and horizontal distribution are combined to allow for independent circulations for various pedestrian flows and avoid mutual interference.

4. New Technologies and New Materials
We have done certain explorations on new technologies and new materials in designing the Guangzhou Asian Games Town Gymnasium. Among them, some are first adopted in China.

5. Conclusion
The construction of a complicated, large-span 3D sports building with continuous curved surface has been a great challenge. Although the Gymnasium is completed within a very pressing timeframe, it has highly accomplished the design scheme. The Project has drawn great attentions and high praise from all walks of the society.

1. 项目背景

项目位于广州亚运城南部，紧邻风景优美的莲花湾，是一个包含体操馆、综合馆、亚运历史展览馆等的综合场馆组群，用地面积 101 086 平方米，总建筑面积 65 315 平方米。经过激烈角逐，广东省建筑设计研究院原创设计的"飘逸彩带"建筑概念设计方案最终获胜并实施。

2. 构思、形式、空间

（1）设计构思：彩带飘逸——表现最具艺术魅力的体育项目——艺术体操。
创作灵感来自广州亚运馆的比赛项目——艺术体操。广州亚运馆的创新设计带出具未来感的前卫建筑形象，彰显了亚运主场馆的独特标志性！
（2）探索新的建筑语言及体验
广州亚运馆展示了一种全新的建筑体验，带来了"步移景异"的戏剧性效果。
（3）营造积极、开放及舒适的城市公共空间
设计了多种形态、层次丰富、连续流动的非均质、半室内灰空间及室外公共空间，最大限度地实现亚运馆的公共性。

3. 功能与流线

各比赛馆的主要功能区均包括：观众区、比赛场地区和内部功能区，流线设计采用了垂直分流与平面分流相结合的方式，使各种人流均有独立的流线，互不干扰。

4. 新技术、新材料

广州亚运馆设计在新技术及新材料应用等方面均作了一些探索，部分为国内首次应用。

5. 结语

对于一个复杂的大跨度三维连续曲面的体育建筑，这是巨大的挑战。但在设计团队的共同努力下，仍然达到了极高的设计完成度，获得了社会各界的高度关注，赢得了赞誉！

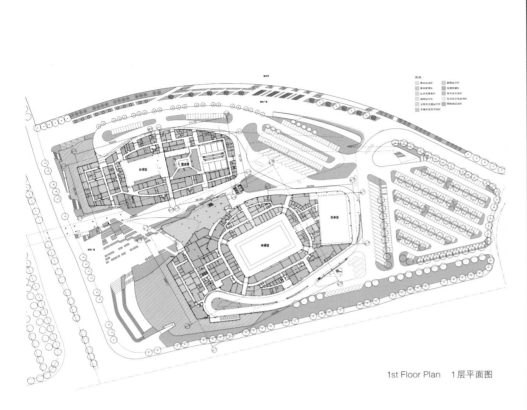

1st Floor Plan　1层平面图

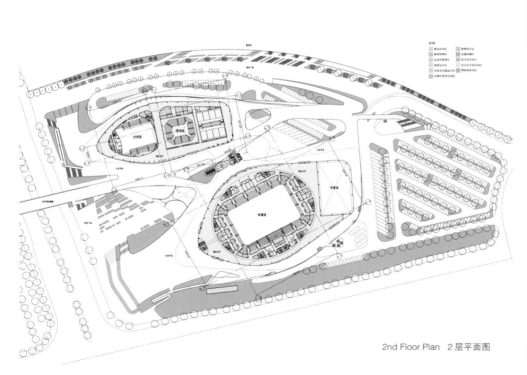

2nd Floor Plan　2层平面图

住宅建筑　RESIDENTIAL BUILDING

The residential building not only is a living space but also interprets the family values. The developers and designers pay more attention to residents' requirements for the environment and the hard and soft decorations within the house. Whether the residential building can inspirit residents' common emotions for the family style living space and create comfortable, healthy, unique environment or not has become the main standards to judge new era residential buildings.

1. Environment-friendly, healthy residential environment and space. Residential environment and space is the first of all factor when people choosing the residence. Space size and environment quality decide the happiness index of the residents who attach lots of importance on the comfortableness and environmental health. Those most expensive probably aren't the most comfortable, while the most healthy living environment is the most important.

2. Community styled, diversified residential building. The modern buildings with various development orientations have get rid of the one-story house form in the before and is developing toward the community and diversification. Therefore, different residence buildings are offered for people to choose. With the procedure of people's changing residential concept, the modern residential buildings gradually transform their forms. And those group residences is easy to be accepted by the market and develop commercially.

The residential buildings expand the spread of technology and reasonable planning of the space. The modern residential buildings' facades develop toward conciseness and diversification and the building functions meet the need of residents to the largest extend.

3. The technology-oriented development is the main trend of group buildings. The reasonable space planning decides the fundamental level of residence. According to the requirements of the group dwelling, the designers use building drawings to satisfy the residents and create comfortable residential blocks to enable people to live happily.

The modern community residence develop towards environmental-friendly, diversification and reasonable planning. No matter what kind of type and property the residential building is, its goal is to provide better service and create healthy life quality. Chinese residence is related with many aspects which decide the residence's future development of China such as national policy, social sustainable development, innovative technology of residence products, finance policy, land policy and so on. In a whole, the residential development trend of China focuses on environmental-protection technology, reasonable space and standard service.

　　住宅建筑不仅仅代表一个居住空间，还承载着居住人群的家庭观念。开发者和设计师也越来越关注居住者对环境的要求和对住宅软硬装的要求。住宅建筑能否激发居住者对家庭式居住空间的共鸣，创造舒适、健康、个性的居住环境，成为衡量新时代居住建筑的主要标准。因此，现代住宅建筑开发了以下一些新的趋势。

　　一、居住环境和居住空间的绿色化、健康化。人们挑选住宅的首要因素就是居住环境和居住空间。空间的大小和环境的好坏决定着家庭居住的幸福指数，居住人群在意的是住宅的舒适程度和环境健康程度。最贵的不一定是最舒适的，最健康的生活环境才是最需要的。

　　二、住宅建筑的社区化和多元化。现代建筑的发展方向多种多样，摒弃了以前传统的平房居住形式，越来越朝着社区化和多元化发展。由此，居住人群也有了各种各样住宅形式的选择。现代住宅建筑随着现代人群对住宅观念改变的进程逐渐转型，适合群体居住的住宅乐于被市场接受和进行商品化发展。

　　三、住宅建筑扩大了科技化和空间的合理规划性。现代住宅建筑的外观，朝着简洁、多样的方面发展，建筑功效也最大限度的为居住人群服务。科技化发展是群体建筑的基本发展趋势。合理规划空间决定着住宅的根本居住水平。设计师们根据群体居住的要求，用建筑绘制成为可以向居住者们可以讲述的语言，打造着舒适程度适合的建筑住宅区，让居住者在居住环境里快乐生活。

　　本世纪，社区住宅的绿色、多元化、合理规划的方向发展。无论是哪种形式和性质的住宅建筑，都是更好地为居住者服务、创造绿色的生活品质。中国的住宅的未来和很多方面相关：国家的政策、社会的可持续发展、住宅产品的技术创新、金融政策、土地政策等，都决定着中国未来居住住宅的发展。但总体而言，中国住宅的主要发展趋势是向着绿色科技化、空间合理化、服务规范化方面发展。

THE RIVERSIDE—EAST PARK
山水文园——东园

Architects: ROGGEO, Beijing New Wra Architectural Design Ltd.
Designer: Roger YU
Location: Beijing, China
Site area: around 350,000 m²
Photographer: Yang Chaoying

设计机构：加拿大诺杰建筑设计事务所、
北京新纪元建筑设计工程设计有限公司
设计师：Roger YU
项目地点：中国北京市
总用地面积：约 350 000 平方米
摄影：杨超英

The Riverside East Park is located on the south of Hong Yan Road, on the west of East Fourth Ring, Chaoyang District, Beijing. It is a 350,000 m² high-end residential community with entertainment and shopping facilities. The Riverside East Park is inspired by the natural and healthy urban life. The building facades adopt the simple and modern German style to achieve the concise and light effect. The materials such as stone, aluminum plate and imitation tile match to each other. The park consists of 35 separate towers and slab-type apartment building. The two-floor commercial space equipped with high-end school and art center is arranged along the Hong Yan road at the north side. The entire project is a good example of the interdependence between living environment and the landscape.

山水文园东园坐落于北京东四环朝阳区弘燕路以南，东四环以西，占地面积约 18 万平方米，是个拥有 35 万平米的高端住宅小区及集休闲购物于一体的商业项目。东园的设计灵感来源于都市中自然、健康的生活，建筑立面采用简约现代的德式风格，追求轻巧明快的效果，建筑材料采用石材、铝板及仿石瓷砖自然搭配。园区由 35 栋独立的塔楼和板楼住宅组成，北面沿弘燕路布置两层商业，同时配有高档的学校和艺术中心等。整个项目很好地体现了居住环境与山水的相互依存。

Elevation 1 立面图 1

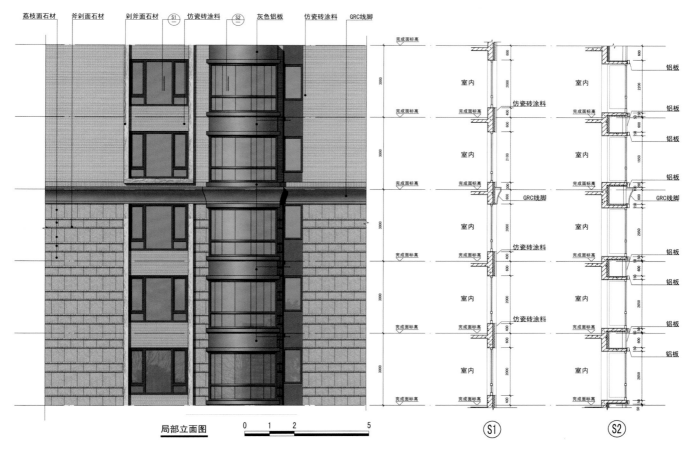

荔枝面石材　斧剁面石材　剁斧面石材　(S1)　仿瓷砖涂料　(S2)　灰色铝板　仿瓷砖涂料　GRC线脚

局部立面图

0　1　2　5

室内　仿瓷砖涂料　铝板

室内　GRC线脚　铝板

室内　仿瓷砖涂料

室内

(S1)　(S2)

Elevation 2　立面图 2

Elevation 1 立面图 1

Elevation 2 立面图 2

SHANGHAI MINI MANSION
上海 MINI 公馆

Architects: Shanghai ZF architectural design co.ltd
Location: Shanghai, China
Site area: 76,600 m²

设计机构：上海中房建筑设计有限公司
项目地点：中国上海市
总用地面积：76 600 平方米

Shanghai MINI Mansion is located on the intersection of Zhenbei Road, Taopu Road and Zhenbei Branch Road, Putuo District, Shanghai City. The site area is 2.4 hectares with the floor area above ground of about 76,600 m², underground construction area of about 19,600 m².

Harmonizing the city relationship, changing the architectural character
The serviced apartment was treated to be twin towers whose connecting line was perpendicular to the elevation road. The supermarket with large volume is located between the two towers This approach can not only coordinate the relationship between the city space and the surrounding buildings, but also make the people feel its iconic effect.

Changing the visual experience, achieving the slimming effect
How to realize the slimming effect visually by design skills on the premise of ensuring the typical floor area unchanging became the important topic of the project. First, the set-back terrace layer by layer on the top of both sides in the main facade of every rectangular tower. Secondly, the main facade uses plenty of white aluminium planes and stones, while the rest uses dark colors.

Improving the construction quality by color of details
The overall building employs black, white and gray colors to chaotic neighboring city context, and reflect the era sense of architecture group. The details pursuit the rhythmic and orderly change, and strive to improve the integrity and quality of the building.

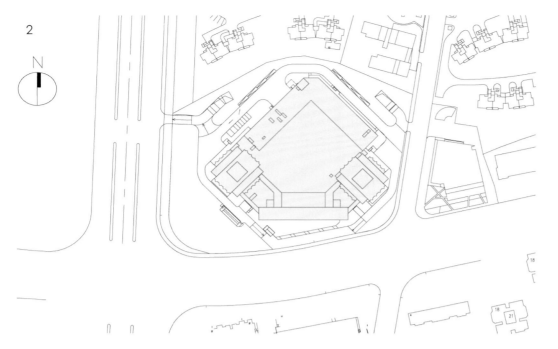

Combining the modeling technique to realize the function of demand
How to arrange hundreds of external units of air-conditionings in facade in order? The vertical lines on the facade resolve the question mentioned above. These vertical lines were treated to be several columns of cages for external units of air-conditionings which enclosed by solid aluminum plane on its front face and shaded by the same color aluminum grille on both sides. The design considers the ventilation, heat dispersion as well as the integrity to achieve the integration of the form and the functions.

Setting up roof garden to extend the space
A large-scale roof garden is set up on the roof of the supermarket to make full use of the space in the case of smaller base.

Plan 平面图

星河世纪城地处上海市普陀区，位于真北路、桃浦路、真北支路的交汇处。项目用地约 2.4 公顷，地上建筑面积约 7.66 万平方米，地下建筑面积约 1.96 万平方米。

协调城市关系 改变建筑个性
设计将酒店公寓处理为双塔，并将其连线垂直于高架布置，大体块的超市位于两塔之间。这样既能较好地协调城市空间和周边建筑的关系，又能使人们感受到其标志性的效果。

改变视觉感受 达到瘦身效果
如何通过设计手法，在确保标准层面积不变的前提下实现建筑视觉上的瘦身，成了本项目设计的重要课题。首先，在每栋正方形塔楼的主立面中利用两侧顶部层层退台处理；其次，主体大量采用白色铝板和石材，其余部分则用深色。

通过细部色彩 提升建筑品质
建筑整体采用黑、白、灰的色彩，以体现建筑群体的时代感。建筑的细部追求具有韵律和次序的变化，力求达到建筑整体性和品质的提高。

结合造型手法 实现功能需求
如何使数百台空调外机在建筑立面上摆脱杂乱无章呢？设计师最终在立面竖线条上找到了解决的办法。设计将这些竖线条处理成一列列空调外机的笼架，在其正面用实心铝板包封，两侧采用同色铝板格栅遮挡。从而使通风散热功能和外观整体性兼顾，实现了造型与功能的统一。

设置屋顶花园 空间得到延伸
设计中利用超市屋顶设置了大面积的屋顶花园。使在基地较小的状况下实现了空间的复合利用。

256

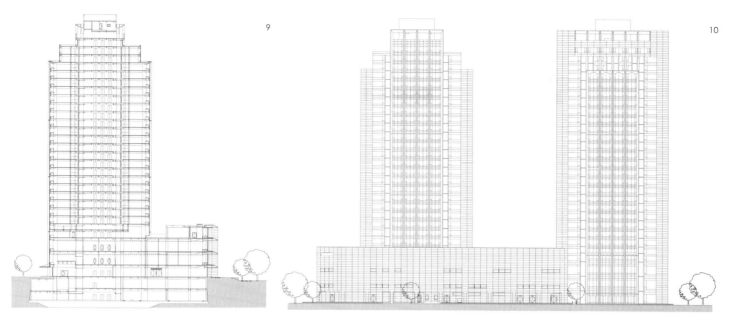

9

10

Elevation　立面图

LONGHU ORIGINAL MOUNTAIN VILLA CLOSE COURT
常州龙湖原山合院别墅

Architects: Shuishi International
Location: Changzhou, Jiangsu, China
Site area: 312,000 m²

设计机构：水石国际
项目地点：江苏省常州市
总用地面积：312 000 平方米

Project Description
In the urban living, the friendly relationship between neighborhood has disappeared, this project focuses on the promotion of communicate between neighbors, meantime, this neighbor relationship could not affect their own living courtyard. Each courtyard is composed of ten households, which creates the entertainment public area. The courtyard aims at letting lonely child find playmates and senior citizens find friends. Public and the private being the opposite, the project has turned hierarchical to interior living space, private courtyard, intermediate public landscape area from inside and outside space, it has been divided into private space, semi-private space and public space. Through the transition of the space, the project utmost solves the contradiction between public and private. The building faced has been designed in Mediterranean architectural style and been composed of two elements of Italy and Spain. Different styles, multiple shapes and luxuriant colors, these kinds of elements make each building a special characteristic and the unique feature of each building outstands.

　　项目介绍
　　现代都市生活中，邻里关系渐渐消失，本案旨在设计一个便于邻里沟通，同时又不影响自家生活的院落。每组院落由十户人家组成，围合出供休闲娱乐的户外公共场所。在这个院落里，让孤独的孩子拥有玩伴，让安享晚年的老人拥有朋友。公共和私密空间相互独立，本案在空间上从内到外，依次递阶为内部的居住空间、自家的小花园、中间的景观公共区，有私密空间、半私密空间以及公共空间三个层级。通过空间的过渡，最大限度解决公共和私密的矛盾。建筑立面上采用地中海建筑风格设计，由意大利和西班牙两种风情元素组成。不同的风格、多样的造型、丰富的色彩使每座建筑拥有不同的表情，彰显个性。

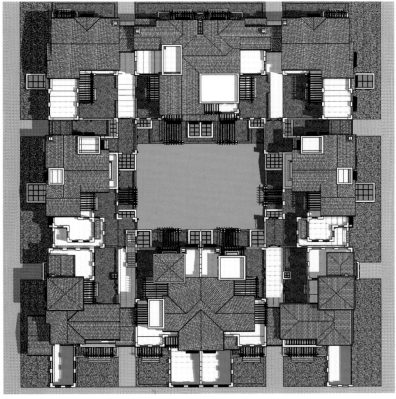

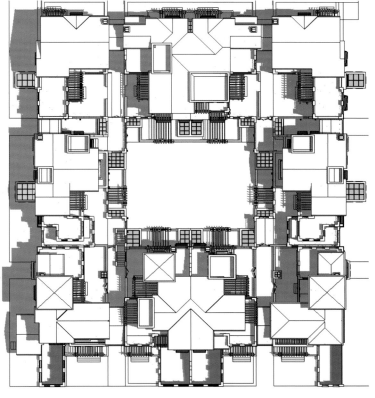

Roof Plan　屋顶平面图

Elevation 1　立面图 1

Elevation 2　立面图 2

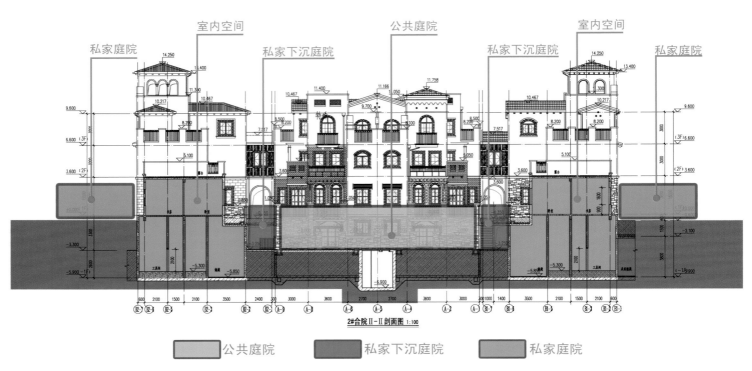

私家庭院　　室内空间　　　　　　　　　　公共庭院　　　　　　　　　室内空间　　　私家庭院

私家下沉庭院　　　　　　　　　　　　　　私家下沉庭院

2#合院Ⅱ-Ⅱ剖面图 1:100

公共庭院　　　　　私家下沉庭院　　　　　私家庭院

Analysis　分析图

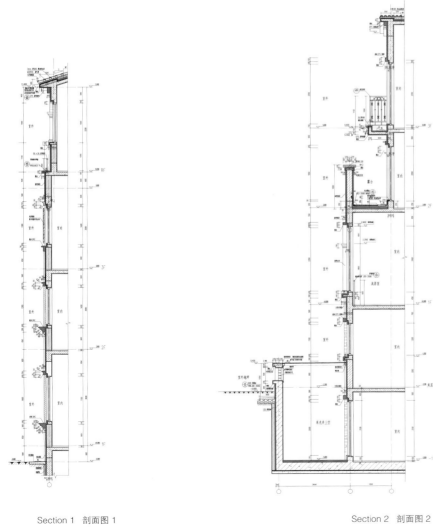

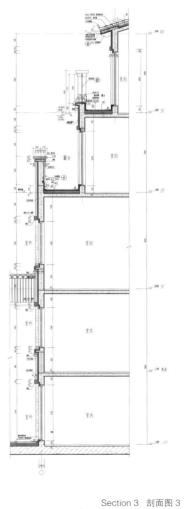

Section 1　剖面图 1

Section 2　剖面图 2

Section 3　剖面图 3

HARBIN SEA ISLAND VILLA
哈尔滨海域岛屿墅

Architects: W&R Group
Location: Harbin, Heilongjiang, China
Site area: 410,000 m²

设计机构：水石国际
项目地点：中国黑龙江省哈尔滨市
总用地面积：410 000 平方米

The design draws on the experience of Shanghai ARTDECO architectural elements to create a villa full of classical aristocratic palace temperament through the heavy, ornate stone, metal and other material language. We try to avoid the mechanical rigid way of the past townhouses' to be in order and array. The 6 townhouses, 8 townhouses, 12 townhouses are just like a single large residence after the re-integration of the overall morphology of townhouses—complete, elegant, grace. There is change in the sequence and combination in the change. Furthermore, the details of courtyard and other buildings' part near to the ground are handled carefully. In the whole design the architecture and environment integrate naturally and harmoniously, which create a high-quality human habitat model.

项目介绍：
设计借鉴海派 ARTDECO 建筑风格元素，通过厚重、华丽的石材、金属等材料语言，希望营造出一种具有古典贵族官邸气质的别墅作品。在整体形态的处理上力求避免以往联排别墅整齐阵列式的机械呆板方式，通过对形体立面的重新整合组织使6联拼、8联拼，甚至12联拼的别墅宛如一幢独立的大住宅——完整、典雅、庄重。在序列中有变化，在变化中有统一，更通过对院落等建筑近地部分的细节处理，真正使建筑与环境浑然一体、和谐自然，创造出高品质的人居建筑典范。

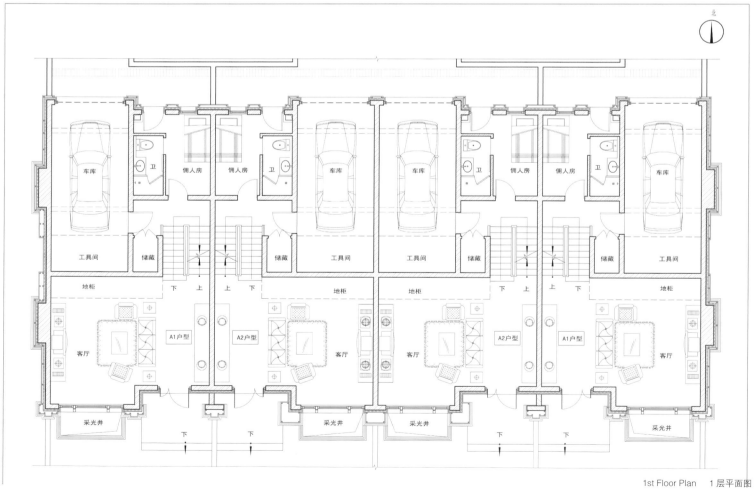

车库　卫　佣人房　佣人房　卫　车库　车库　卫　佣人房　佣人房　卫　车库

工具间　储藏　下　上　上　下　储藏　工具间　工具间　储藏　下　上　上　下　储藏　工具间

地柜　地柜　地柜　地柜

客厅　A1户型　A2户型　客厅　客厅　A2户型　A1户型　客厅

采光井　下　下　采光井　采光井　下　下　采光井

1st Floor Plan　1层平面图

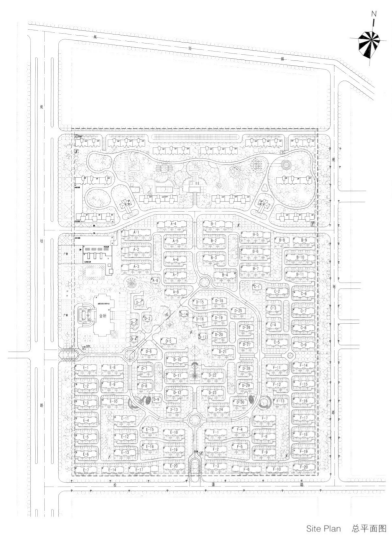

Site Plan   总平面图

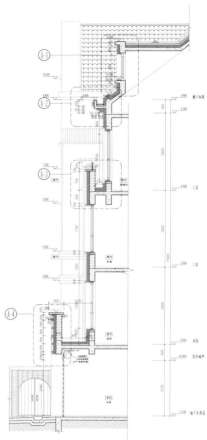

Detail 1 细节图 1

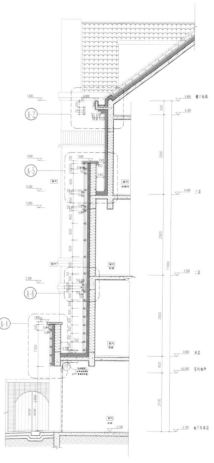

Detail 2 细节图 2

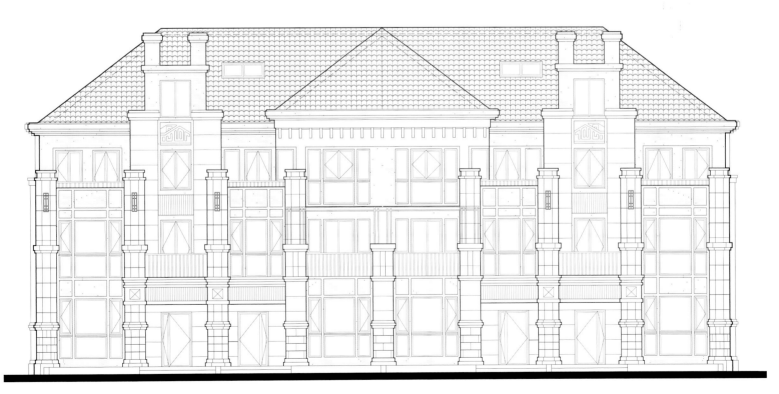

South Elevation   南立面图

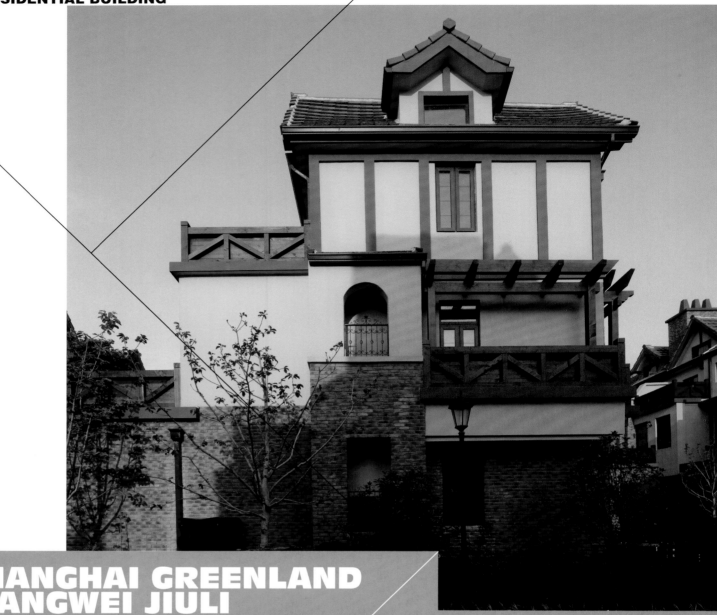

SHANGHAI GREENLAND QIANGWEI JIULI

上海绿地蔷薇九里

Achitects: W&P Gronp
Location: Shanghai, China
Site area: 299, 814 m²

设计机构：水石国际
项目地点：中国上海市
总用地面积：299 814 平方米

Shanghai Greenland Qiangwei Jiuli is located in the intersection of Guangfulin, Jiasong South Road, new town Songjiang. The project is committed to create a residential area with typical British style which includes townhouses, high-rises, commercial streets and squares. The high-rise and low-rise dwellings combine into residential areas respectively. And the high-rise areas break through the traditional closed neighborhood level to naturally transit from public space to private space at the scale of walking distance. The project makes some trials to put economical houses in low-rise residential areas which employ two models, 75 m² and 95 m², to create townhouses with detached single-family style. The marketing slogan of the project is that urban white-collar workers can afford to purchase villas 10 years early. In the model of 95 m², it includes many value-added spaces such as a living room with a width of 5.1 m, three bedrooms, separate parking, attached basement, loft and terrace. Householders can enjoy an indoor space of 130 m² and an outdoor private space of 45 m² with only paying a floor space sold of 95 m². In this way, they will experience the villa life in a limited economical area.

上海绿地蔷薇九里位于上海市松江新城嘉松南路广富林路口，该项目致力于打造一座集原味英式联排别墅、高层、商业街、广场于一体的纯正英伦风格居住区。整个布局由高层与低层住宅分别成区，其高层社区打破传统封闭的邻里层次，以步行距离为尺度，从公共空间自然过渡到私密空间。本项目在低层住宅产品当中尝试了经济户型的可能，采用75平方米与95平方米两个面积来设计独门独户式联排住宅。项目的营销口号是：让城市白领提前十年住上别墅！在95平方米的面积当中，设计5.1米面宽的客厅、三个卧室、独立车位，赠送地下室、阁楼、住宅、露台等增值空间。以95米的销售面积享受130平方米的室内空间与45平方米的户外私家空间，真正以经济的面积体验别墅的生活品质！

Elevation　立面图

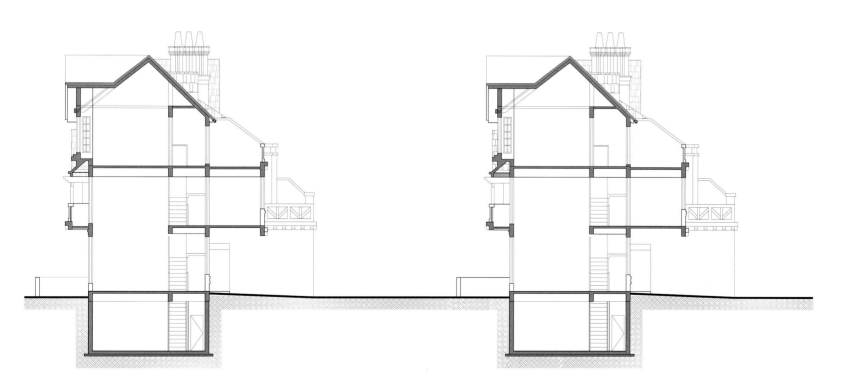

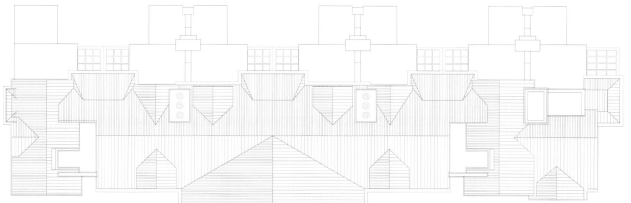

Roof Plan 屋顶平面图

HOP KING · FRAGRANT SEASONS
合景 · 香悦四季

Architects: Beijing Victory Star Architectural & Civil Engineering Design CO,.LTD
Location: Beijing, China
Site area: 600,000 m²

设计机构：北京维拓时代建筑设计有限公司
项目地点：中国北京市
总用地面积：600 000 平方米

Project Summary
Hop King• Fragrant Seasons is located close to Olympic Water Garden, Shunyi District. The total building area is 600,000m². The project consists of luxuries modern houses, villas, theme-oriented commercial street, serviced apartment and other kinds of buildings. The architects attempt to create a livable area integrated with green environment, characteristic waterscape, featured street, demonstration new cityscape and harmonious living.

1. Plot Ratio and Building Density
With a planning plot ratio of 1.1, 45 m building height limitation, this project can be called as a "low plot ratio" community. he project was proposed to be "low architectural density and high-quality of landscape".
2. Planning Structure
(1)Landscape System
In order to create an integrative regional landscape system, the central green belt connect the city green spaces and at lost combined all of them to form a large vibrant community landscape system.
(2)Transport System
In the east and west plot, the transport system was consist of outer ring road and inter cross road. The entrances and road lines for motor vehicle and people are independent.
(3)Layout
In the plan, all kinds of residential buildings keep together and without any separation.

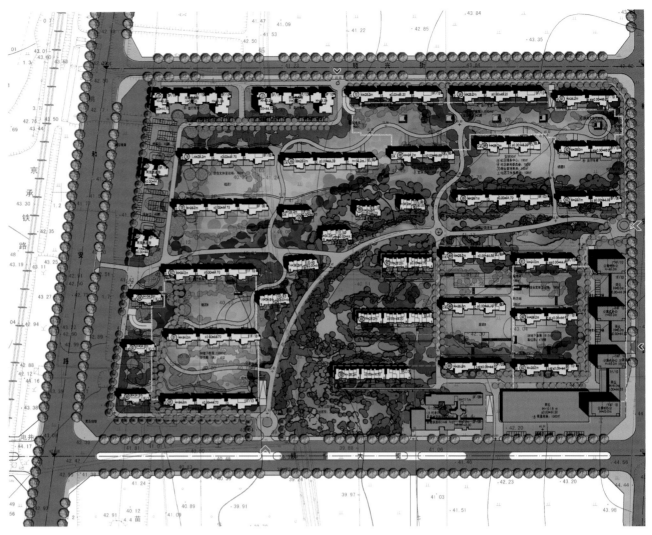

Site Plan 1 总平面图 1

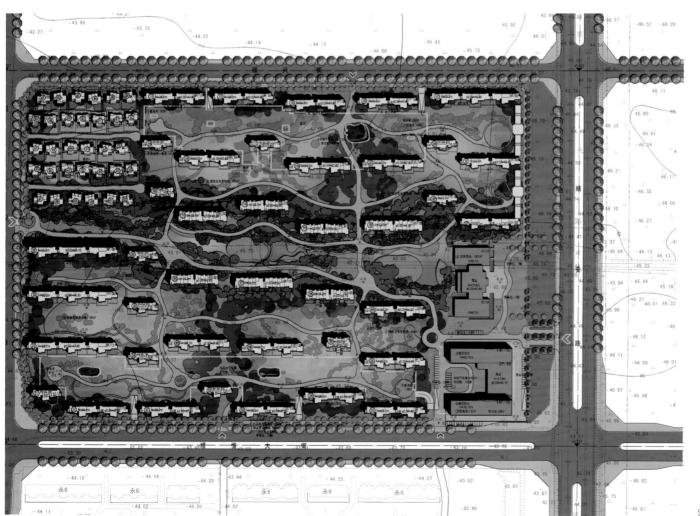

Site Plan 2 总平面图 2

3. Single building type

In this project, most of the single residential buildings are 9-11 stories. And the house type can be generally classified into small size (one elevator three family), big size (one elevator two family), villas and other types. Various housing types are favorable to a harmonious coexistence environment of residents.

4. Facade image

All the facades were designed to be the Art-Deco style, which makes the building looking tall and well-proportioned, unified decorative designs of building top. The use of pale yellow and brown texture coating in facade was on account of the cost and an impressive building appearance.

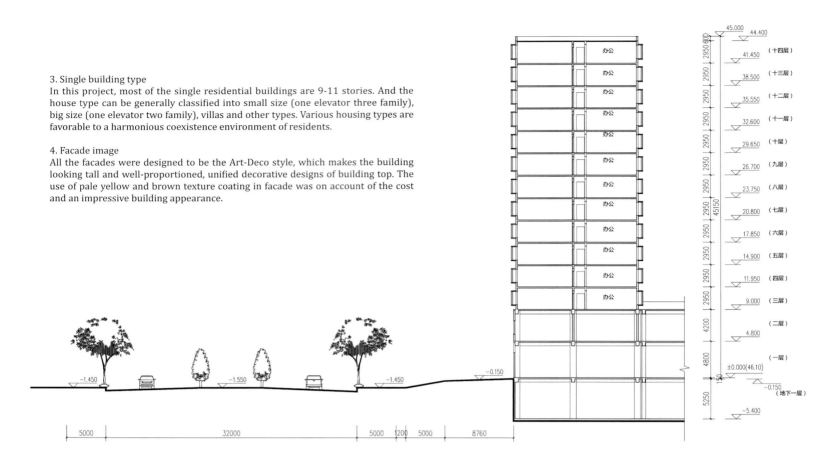

Elevation 1　立面图 1

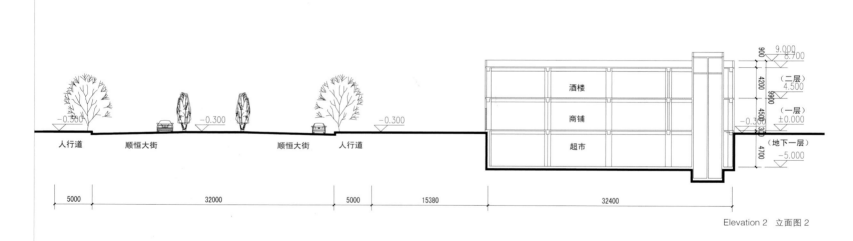

Elevation 2　立面图 2

项目概况

合景·香悦四季坐踞顺义奥运水上公园畔，建筑面积约 60 万平方米，涵纳洋房、别墅、主题商业街、酒店式公寓等多元建筑形态。倾力打造一座集"绿色生态、特色水景、国际品质、主题商街、示范新城、和谐生活"为一体的宜居住区。

规划特色

1. 容积率和建筑密度

本项目用地规划容积率为 1.1，限高 45 米，属于"低容积率"社区。在"低容积率"条件下，同时实现"低建筑密度、高绿化景观品质"。

2. 规划结构

景观系统设计力图创造一种融合的区域景观体系，将城市公共绿地和社区内景观绿化充分结合，有机互补，形成一个充满活力的城市区域景观绿化体系。

2) 交通系统

东西两区分别采用外环加十字形联系道路的区内路网体系，将区内交通充分联系起来。机动车和人行的入口及流线都在最大程度上相对独立，形成安全有效的人车分流系统。

3) 产品分布

在规划设计中，住宅类型多样化，你中有我，我中有你，没有明显的区域等级区隔。

3. 单体户型设计

本项目住宅单体设计大部分采用 9 至 11 层小高层形式。分为一梯三户小户型和一梯两户大户型以及别墅等多种类型，追求住宅产品的多样化，有利于社区内各众人群的和谐共生。

4. 形象设计

本方案立面形象统一采用 Art-Deco 风格，使建筑呈现高挑匀称的飘逸风格，同时强调建筑顶部装饰处理。外墙采用咖啡色及浅黄色质感涂料，强调线条细节比例，有效控制成本并保证建筑外观效果。

QINGYUAN YUNSHAN SHIYI
清远云山诗意

Architects: Shing Partners
Location: Qingyuan, China
Site area: 500,000 m²

设计机构：汉森伯盛国际设计集团
项目地点：中国清远市
总用地面积：500 000 平方米

Qingyuan Yunshan Shiyi project continues and improves the brand of Yunshan Shiyi Renjia. The design combines more modern and warm elements to gain unanimous praise. The achievement of 30% unit-price higher than the surrounding property proves once again that "good design brings high additional value". The project aims to create a high-grade residential community with low FAR, low density and high landscaping ratio, which is based on modern Huizhou-style architecture and combined oriental garden element.

About architectural style, the project has inherited and improved the oriental features and high quality of Yunshan Shiyi in Guangzhou. In this Huizhou-style architecture, the modern technologies, color techniques and decorative materials bring modern and simple atmosphere, creating an elegant, modern, distinguished high-grade community.

As to landscape design, 80,000 m² large oriental charming landscape gardens around the project was built. Qingyuan Yunshan Shiyi is themed by water, enjoying laminated water, small stream and artificial lake. Through different forms of water, combined with landscape architectures, stones, screen wall, an interesting Chinese-style garden emerges.

The supporting facilities are complete, the entrance axis is of beautiful landscape, and the commercial space is creative and impressive. The clubhouses, swimming pool, tennis court, badminton court, gymnasium, reading room, sinology museum provide the residents with the communication platform.

Qingyuan Yunshan Shiyi combines modern technology with the traditional architectural styles organically to create a harmonious residence new concept.

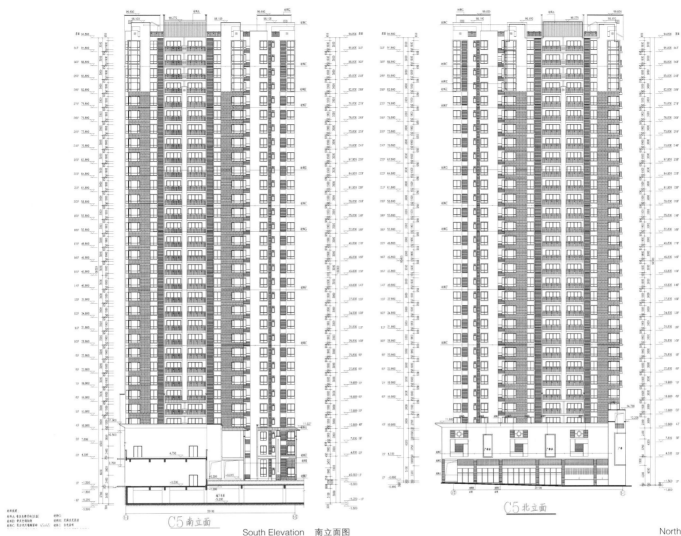

South Elevation　南立面图

North　Elevation　北立面图

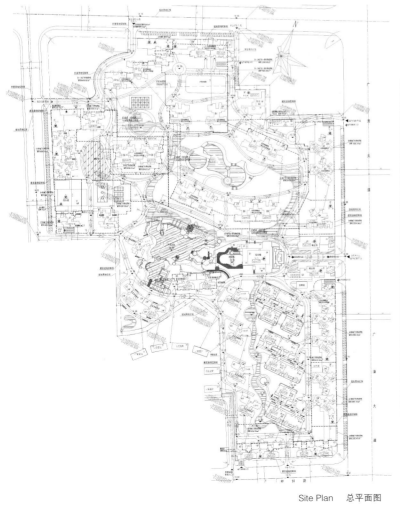

Site Plan　总平面图

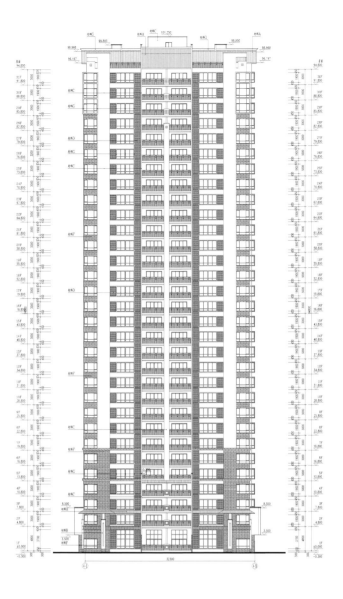

Elevation　立面图

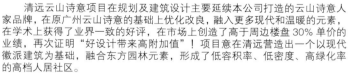

　　清远云山诗意项目在规划及建筑设计主要延续本公司打造的云山诗意人家品牌，在原广州云山诗意的基础上优化改良，融入更多现代和温暖的元素，在学术上获得了业界一致的好评，在市场上创造了高于周边楼盘30%单价的业绩，再次证明"好设计带来高附加值"！项目意在清远营造出一个以现代徽派建筑为基础，融合东方园林元素，形成了低容积率、低密度、高绿化率的高档人居社区。

　　在建筑风格风面，清远云山诗意继承了广州"云山诗意"系列的东方特色及优秀品质，同时又非常注重创新。在徽派建筑的设计风格中，通过现代技术手段、色彩技巧和装饰材料，赋予传统徽派建筑现代且简约的气息，使之成为优雅、现代、尊贵的高档社区。

　　在园林设计方面，8万平方米超大东方神韵环区山水园林，清远云山诗意整体以"水"为主题，叠级流水、涓涓小涧、人工湖泊，通过不同形态、不同类别的水体，结合精心雕琢的景观建筑、奇石、影壁等，形成奇趣盎然的中式园林核心景观。

　　小区配套齐全，入口轴线景观优美，商业完善且设计有创意，令人印象深刻。会所、游泳池、网球场、羽毛球场、健身房、阅读室、国学馆等设施，为业主提供丰富业余生活的交流平台。

　　清远云山诗意，将现代的技术和传统的建筑风格有机结合，创造了现代、建筑、文化三位一体的和谐人居新理念。

THE WORLD ARCHITECTURAL
FIRM SELECTION | 287

综合建筑　COMPLEX BUILDING

After the era that the distinction between the residential real estate and commercial real estate is strict,the single real estate development mode could no longer satisfy the residency requirements of people's lives,also could not adapt to the rapid development of the real estate.Creating a new real estate development mode has become a new issue that developers need to facing. Mixed-using building integrates many functions of the city, including commerce, office, residence, hotel, exhibition, catering, conference, entertainment and transportation, and establishes a dynamic relationship of interdependence in various parts,then forming a versatile, high-efficiency building community. As the urban mixed-using building has all functions of modern city, they are often called as "city within a city".

A successful mixed-using building need to complete with four points.First of all, it needs to be creative and correspond with the external shape of the era aesthetics. Secondly, it needs to make good mixed-using planning, scientific and reasonable arrangement of residence, hotel-style apartments, office buildings, shopping malls, hotels, theme parks and other projects. However, these depend on the local urban development process, family income, residents' purchasing power and consumption structure. Thirdly, it needs to reasonably arrange the stream of people, traffic organization, and combination of space, etc. Different industries do not interfere with each other, making complex really play a compact, comprehensive role. Finally, it needs to pay attention to the combination of architecture and ecology, to increase the green area and achieve humanistic ecology. So all kinds of people in the complex could live together with each other harmoniously.

Mixed-using buildings are not only the landmark buildings of city, but also have been the standard of international life style system of each big city's commercial center district.They are the symbol of high quality city life, they can improve the city's overall image quality and value.The complex of mixex-using building determines that it has strong social function which is the engine of regional economy and the main factors that could enhance urban economy and culture. Excellent commercial complex can enhance the flow of the city, improve the quality of life and consumption,and become an important tourist attraction in the city. All these can enhance the attraction of the city.

The development of the modern city is transformed from extensive to intensive direction. Urban mixed-using buildings are more and more emphasizing on the city's openness and integration, focusing on the construction of urban public space, which forms city's public spaces, such as the transportation hub, cultural square.Through the linkage with city, regional development has been integrated.

　　经历了将住宅地产和商业地产严格区分的时代后，单一的地产开发模式已经渐渐不能满足人们对生活居住的要求，也逐渐跟不上地产业的迅速发展，因此，崭新的城市综合体开发模式已成为地产商们的新型课题。

　　城市综合体是指将城市中的商业、办公、居住、旅店、展览、餐饮、会议、文娱和交通等生活功能融为一体，并使各部分间相互依存、相互受益，从而形成的一种多功能、高效率的建筑群落。由于城市综合体基本具备了现代城市的全部功能，因此又被称为"城中之城"。

　　成功的城市综合体的发展有几点必须涵盖：首先是独具的创意，符合时代审美观的外在形态；其次，好的规划设计，包括科学合理地安排住宅、酒店式公寓、写字楼，也涵盖商场、酒店、主题公园等多种项目，而这些都要根据当地城市发展进程、家庭收入情况、居民的购买力和消费结构来决定；然后，要做好人流组织、交通组织、空间组合等，不同业态之间要做到互相不干扰，使综合体真正发挥紧凑、综合的功能；最后，注重建筑与生态相结合，增加绿化面积，实现人文生态，使综合体中的各种人群和谐相处。

　　这种城市地标性建筑，正逐渐成为各大城市商业中心区和国际化标准的生活模板，是高品质城市生活的标志；它提升着城市的价值、品质和整体形象，是拉动区域经济和文化的引擎。它的多重复合性决定了其具有强而多的社会功能——优秀的商业综合体能够增强城市的流通力，提高居住人群的生活和消费品质，也能成为地方重要的旅游景点，对于增强城市吸引力有重要作用。其间最主要的一点是立体交通的配合。城市综合体的规划必须与交通等规划设计相协调，交通承载着其发展的重要疏通作用。打造完善的城市公共空间，必然需要稳定流畅的交通枢纽和精神丰富的文化广场等。交通与城市、区域形成联动发展，实现综合体一体化。

　　现代城市的发展正由粗放型向集约型方向转化，城市综合体越来越多地强调开放性与融合性。日趋完备的商业办公、衣食、交通和文化氛围，促进着多功能区的融合。城市综合体之间相互促进，形成群体城市发展的大好形式，整体城市覆盖率的大幅度提升指日可待。

WUHAN TOWER
武汉塔

Architects: Dot-A Ltd.
Architectural design & Research Institute
Project architect : Duo Ning
Designer team: Chang Qiang, Gao Yan, Huang Di, Lu Yangying,
Qin Tao, Kan Zhuowei
Location: Wuhan, China
Site area: 27,768 m^2
Gross floor area: 203,066 m^2

设计机构：度态建筑事务所
合作单位：清华大学建筑设计研究院
项目负责：朵宁
设计团队：常强、高岩、 黄荻、路阳英、
覃韬、阚卓威
项目地点：中国武汉市
总用地面积：27 768 平方米
总建筑面积：203 066 平方米

Hubei Transport Investment Headquarters is a mix-used complex project located in the new CBD area of Hubei. The project comprises two major parts, the Headquarter and a series of residential neighborhood. Composed of office, hotel, specialized services and business center, the Headquarters was expected to be a new icon of the new district, which pronounced our main design focus.

In this project, we questioned the fundamental mechanism for iconic effect of skyscraper instead of formal approach, i.e. extracting formation logic by analyzing and understanding the nature of Hubei Transport Investment Co., which is responsible for design, constructing and administrating transportation infrastructure. The concept was then to create a new type of spatial organization with the circulatory principles borrowed from transport infrastructure to inform the entire tower from horizontal to vertical, in the meantime, to fulfill the independence and connection between multiple programs.

The final design concept was inspired by clover-leaf interchange, which was employed as four intertwined legs stretching around to form four courtyards in response to the four main functional zones, and slowly winding up to merge into the four vertical towers, which finally joined as one unity at the top.

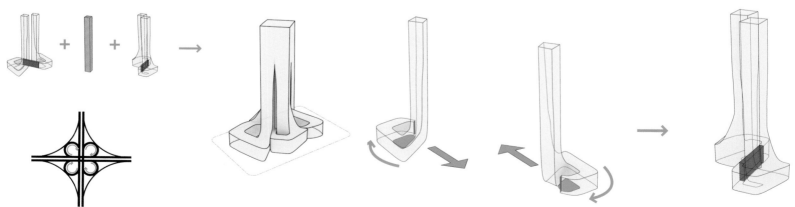

Concept Map 概念图

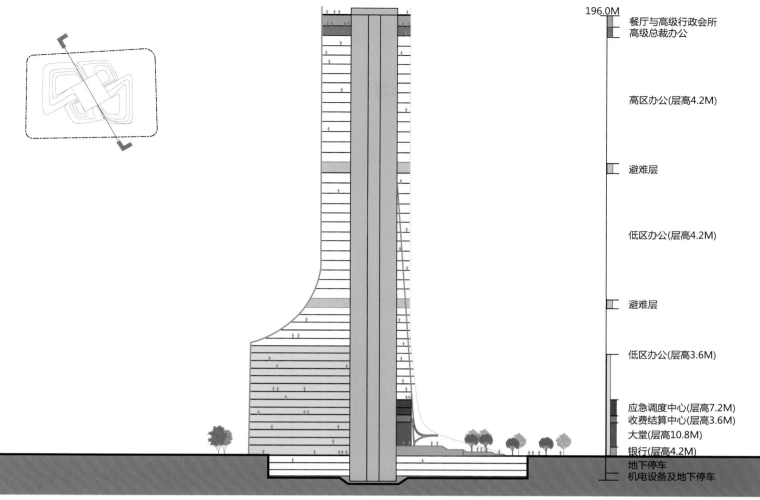

196.0M

餐厅与高级行政会所
高级总裁办公

高区办公(层高4.2M)

避难层

低区办公(层高4.2M)

避难层

低区办公(层高3.6M)

应急调度中心(层高7.2M)
收费结算中心(层高3.6M)
大堂(层高10.8M)
银行(层高4.2M)

地下停车
机电设备及地下停车

Section   剖面图

湖北交投是一个集集团总部和住宅为一体的大型城市综合开发项目。我们的设计核心是总部大楼，它主要由办公、酒店、业务服务、对外营业四个功能块组成，位于武汉新城区的主要干道，将成为该区域的标志性建筑之一。

根据项目具体任务书的要求和湖北交通投资集团的企业性质，我们设计的策略是从这栋摩天楼的最原始的组成机制——交通流线出发，意在通过参考交通设施的流线逻辑，自然导出一个新的摩天楼的形式，它同时可以有效满足功能区之间既分隔又联系的使用需求。

方案最后的形式生成过程从首蓿叶立交桥中受到启发，把建筑的底部组织成由四个条状体量盘旋交织的平面流线关系，在地面层围合成四个院落，并有机地升成摩天楼的主体部分，策略性的暴露内部的交通核心，强化交通的设计概念。

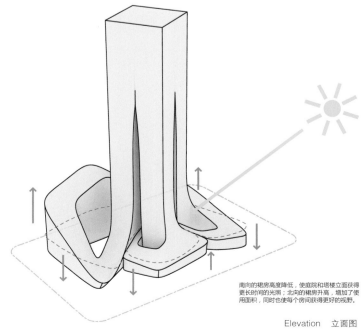

南向的裙房高度降低，使庭院和塔楼立面获得更长时间的光照；北向的裙房升高，增加了使用面积，同时也使每个房间获得更好的视野。

Elevation   立面图

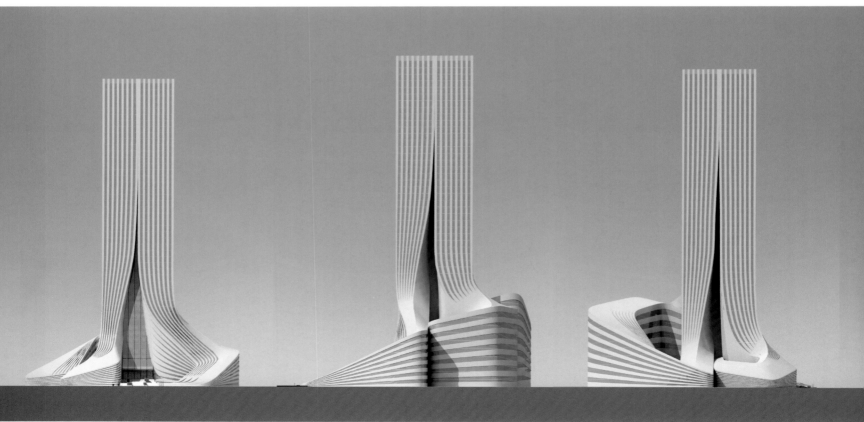

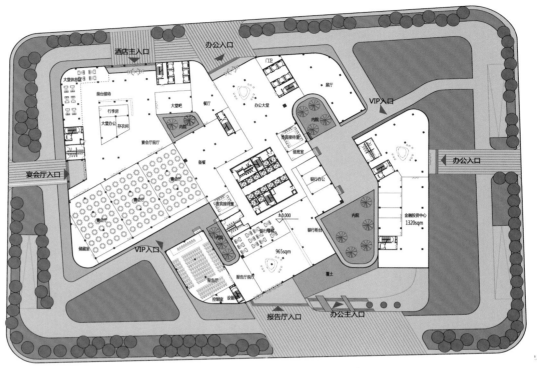

1st Floor Plan　首层平面图

Modelling　模型图

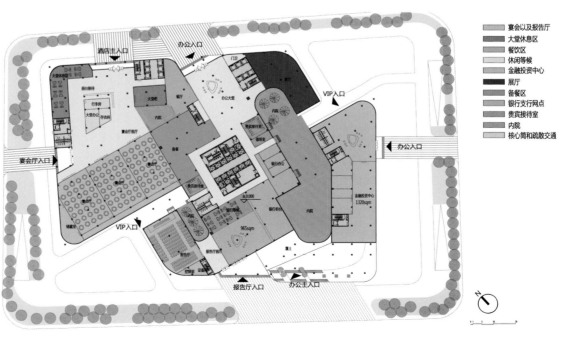

夦宴会以及报告厅
大堂休息区
餐饮区
休闲等候
金融投资中心
展厅
备餐区
银行支行网点
贵宾接待室
内院
核心筒和疏散交通

1st Floor Analysis　首层分析图

THE WINDOW OF GREENLAND, ZHENGZHOU
郑州绿地之窗城市综合体

Architects: Pan-Pacific Design Group Ltd. (Canada)
Location: Zhenzhou, China
Site area: 51, 620 m²
Gross floor area: 482, 600 m²

设计机构：泛太平洋设计集团（加拿大）
项目地点：中国郑州市
总用地面积：51 620 平方米
总建筑面积：482 600 平方米

The Window of Greenland, Zhengzhou is located in the west square of Zhengzhou comprehensive transportation area, where the east is Zhengzhou New Station, the north is Qili River and the south is station green square. Based on the strong economic influence of high-speed rail transportation hub, effect of pedestrian flow and convenient transportation, this area becomes to large urban mixed-use area. The mixed-use area combines commercial office, tour, residential and entertainment functions, including commercial, business, hotel and other industries, these industries supplement each other and compound symbiosis, creating a fully functional, an efficiency use of land and space, a high concentration of cultural content integrated blocks. The design of this project uses a lot of liner elements with some arc elements. The shape is concise but lively, simple and practical without losing the magnificence. The commercial lines are clear and reasonable, commercial nodes are varied in size that echoed the crowd and occupied a proper position. According to the function of Commercial Street scale and combined the needs of underground business, the design was made orderly and space pleasant.

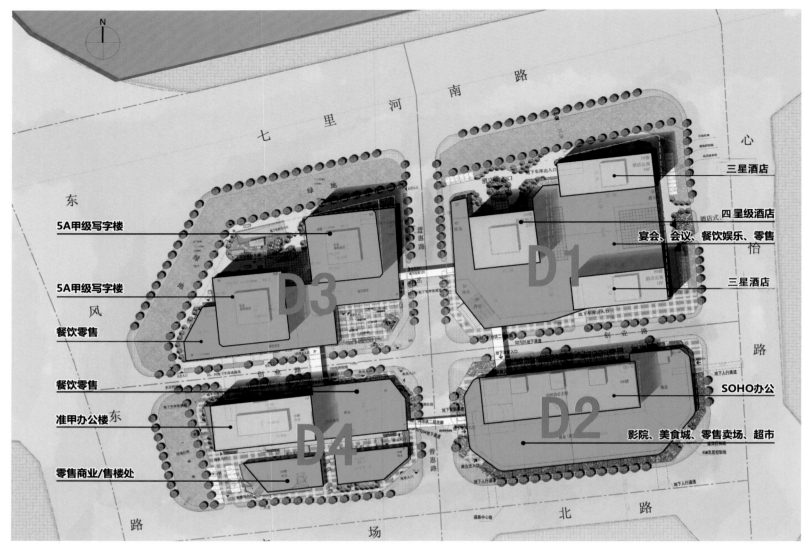

三星酒店

四星级酒店

宴会、会议、餐饮娱乐、零售

三星酒店

5A甲级写字楼

5A甲级写字楼

餐饮零售

餐饮零售

准甲办公楼

零售商业/售楼处

SOHO办公

影院、美食城、零售卖场、超市

D1

D2

D3

D4

Site Plan　总平面图

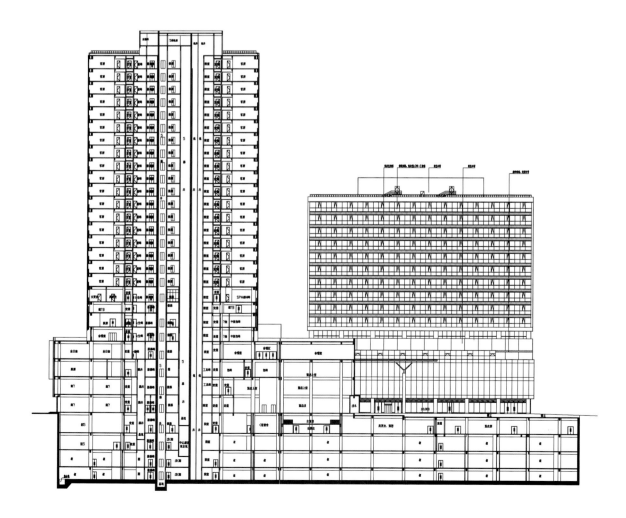

Section　剖面图

绿地之窗项目位于郑州综合交通枢纽核心区的西广场，基地东侧为郑州新客站，北侧为七里河，南侧为站前绿化广场。它是依托于高铁交通枢纽强大的经济影响力、人流汇聚力和交通便捷性形成的大型城市综合体。该综合体集合商务办公、旅游、居住、文化娱乐等功能于一体，业态涵盖商业、办公、酒店等，多种业态相辅相成，复合共生，共同形成一个城市功能齐全、土地和空间高效利用、文化含量高度集中的综合街区。本案的设计采用大量的直线元素，其间穿插部分弧线元素，造型简洁而不失活泼，简单实用而不失大气，商业流线清晰合理，商业节点大小错落，呼应人流，位置得当。商业街尺度根据功能结合地下商业的需要，设计得收放有序，空间宜人。

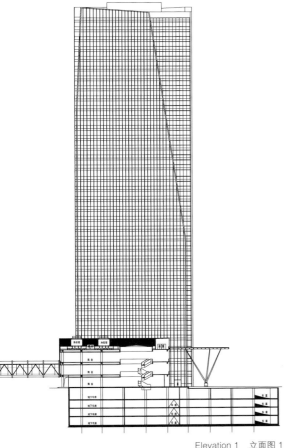

Elevation 1　立面图 1

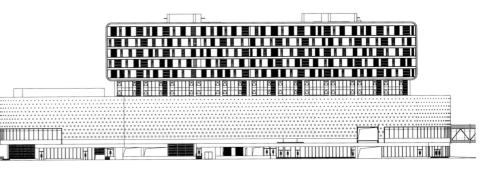

Elevation 2　立面图 2

THE COSMOPOLITAN RESORT CASINO

国际赌城度假村

Architects: Arquitectonica
Associate architect: Friedmutter Group
Interior designers: Dougall Design Associates, Inc.
　　　　　　　　　 Paul Duesing Partners
Location: Las Vegas, Nevada, USA
Site area: 618,400 m²
Photographer: Rick Fowler

设计机构：Arquitectonica
合作设计：Friedmutter Group
室内设计：Dougall Design Associates, Inc.
　　　　　 Paul Duesing Partners
项目地点：美国内华达州拉斯维加斯市
总用地面积：618 400 平方米
摄影：Rick Fowler

Project Description
A 3,016 unit condo/hotel in two towers (48 and 47 story) over a 5-story podium containing a casino, retail, restaurants, 1,800-seat theater, rooftop club, pools and spa, underground parking.

Design Intent
The building occupies the entire site like in a major urban setting and engages the sidewalk, populating it with store fronts that beckon to tourists by its transparency. The glass podium engages the street while its panels fold theatrically in dramatic angles. Three levels of shops and restaurants within become part of the boulevard experience. The architecture is not based on a place or period in history but is instead designed as modern glass sculptures. They create forms that are abstract. These glass prisms twist and fold and define the skyline.
The roof of the podium established a whole new environment. The entire roof is covered in a layer of water. Islands dot this sea, and beaches surround it to create a special world. Two sculptural 50-story towers emerge from this water surface and their form gets reflected in it adding to the drama of their forms. Their glass facades glisten in the night sky with comets speeding toward the boulevard. The buildings are no longer static but instead in motion. They announce the casino and hotel by their sheer presence. They are the marquee.

项目概括
　　该项目由两座 48 层和 47 层的大厦组成，共有 3016 套公寓和酒店房间，超过 5 层楼的高台里设有赌场，零售店，餐厅，容纳 1 800 人的剧院，屋顶天台俱乐部，泳池 spa 和地下停车场。

设计意图
　　这一建筑占据了市区的主要地块，建筑面向街道的商业店面采用了对游客充满诱惑力的透明橱窗设计。玻璃建筑与街道相接的侧翼部分的转折效果富有艺术美感。三层商铺和餐饮店也已融入街景。这一建筑已经不再只是属于某个地方或者某个时期，而是成为了一座现代性玻璃雕塑。建筑创造出了一种抽象的形式艺术。玻璃棱镜缠绕折叠，刻画出了独特的天际脉络。
　　玻璃建筑的顶部又是另一番全新的景象。整个楼顶被水层覆盖。就像是一个由小岛点缀着，海岸环绕着独特的海洋世界。两座巍然屹立的 50 层大楼从水面浮出，大楼的外形倒影在水面，形成另一层次的戏剧性的艺术效果形态。大楼的玻璃幕墙在夜空中闪烁的光亮像是彗星在街道上驰骋而过，整个建筑不再静止，转而变幻万端。建筑绝妙的外在形态高效地展示出了赌场和酒店的存在感。建筑是美好的承载，独特的华盖。

HONGQIAO COMPREHENSIVE TRAFFIC HUB
虹桥综合交通枢纽

Architects: ECADI
Location: Shanghai, China
Site area: 26,260,000 m²
Gross floor area: 1,420,000 m²

设计机构：华东建筑设计研究院
项目地点：中国上海市
总用地面积：26 260 000 平方米
总建筑面积：1 420 000 平方米

West Terminal and East Traffic Square of Hongqiao Comprehensive Transportation Hub Shanghai Hongqiao comprehensive transportation hub is a landmark city infrastructure construction combined functionality, networking, pivotal role in one, a assemble district to speed up the development of modern service industry in Shanghai, a crucial project to provide full service for the Yangtze River Delta and even the whole country. The hub combines aviation, high-speed rail and light rail, highway, maglev train, metro, bus and taxi in one, which sets a precedent for multiple transportation ways with zero transfer, enabling fast gathering and distribution of stream of people in a large district. It is a unique, world-class modern transportation hub.

The west terminal of Hongqiao airport is an important part of the hub, which takes "functionality is landmark" as the starting point, "people-oriented" as the design concept, "sincerity and pragmatism" as the architecture origin, emphasizing clear functions and convenient process to provide high quality services for passengers; coordinating the ratio scale, style and details on the shape and taking "Rainbow" and "bridge" as the motif, the design reflects the place features of the architecture and interprets the simple, crisp, modern characteristics of traffic building, which combines the function and form perfectly. The west terminal of Hongqiao airport has a building area of 3,640,000 m² and 21 million annual throughputs of passengers (long-term 30 million), whose efficient, introverted design thinking plays a significant role in the process of today's rapid urbanism development.

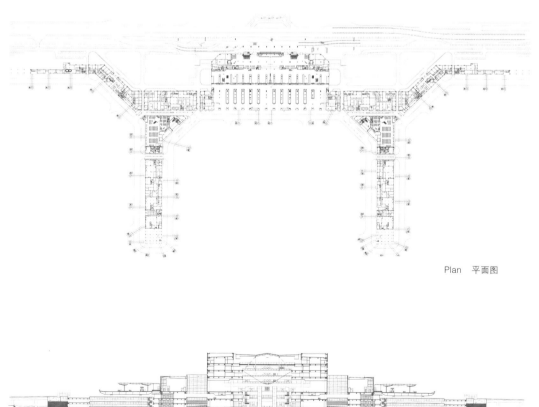

Plan 平面图

虹桥西航站楼和东交通广场构成了上海虹桥区的综合交通枢纽，它们是上海功能性、网络化、枢纽型城市基础设施建设的标志性工程，是上海加快现代服务业发展的集聚区，是上海服务长三角、辐射全国的重大工程。枢纽集航空、高铁城铁、公路、磁浮、地铁、公交出租等功能于一体，开创了多种交通方式零换乘的先例，实现了跨区域、大范围人流的快速集散，是独一无二的世界级现代化交通枢纽。

虹桥机场西航站楼是枢纽的重要组成部分，以"功能性即标志性"为出发点，以"以人为本"的设计理念，以"诚恳务实"的创作思想回归建筑的本源，在设计上强调功能清晰、流程便捷，为旅客提供高品质的服务；在造型上通过比例尺度、风格和细节的协调，以"虹"和"桥"为母题，体现建筑的场所特征，诠释交通建筑简洁、明快、充满现代感的性格特点，做到功能与形式的完美统一。虹桥机场西航站楼建筑面积 3 640 000 平方米，年旅客吞吐量 2 100 万（远期 3 000 万），其高效、内敛的设计思想，在当今城市化快速推进的进程中具有非凡的意义。

Elevation 立面图

PARK VENTURES — THE ECOPLEX ON WITTHAYU
Ventures 花园

Architects: P&T Group
Location: Bangkok, Thailand
Site area: 81, 450 m²
Photography: P&T Group

设计机构：巴马丹拿集团
项目地点：泰国曼谷市
总用地面积：81 450 平方米
摄影：巴马丹拿集团

The Park Ventures Ecoplex on Witthayu is located within a network of Bangkok's main tree-lined arteries connecting urban fabric with nature. The eco-conscious mixed-use complex rises from a park-like environment creating an oasis within the centre of Bangkok's central business district. A prominent sculpture that pays homage to the traditional Thai "Wai" symbol, it is located at the corner of a dynamic road crossing. The building's North-South orientation proposes both a landmark as well as an advanced construction which minimizes energy consumption by means of its strategically placed solar orientation.

Park Ventures is located on a site measuring 8,145 m², rising an impressive 33 storeys above ground with 1 ½ levels of basement. It encompasses a total building area of 81,450 m². whilst providing 13 floors of commercial office space and a 5-star luxury hotel by international standards boasting 242 rooms. The project provides a total number of 580 car parking spaces.

The mixed-use complex houses large column-free office space within the lower portion of the tower. A traditional central-serviced core design allows for rectangular shaped tenancy divisions, promoting flexible tenant layouts and arrangements.

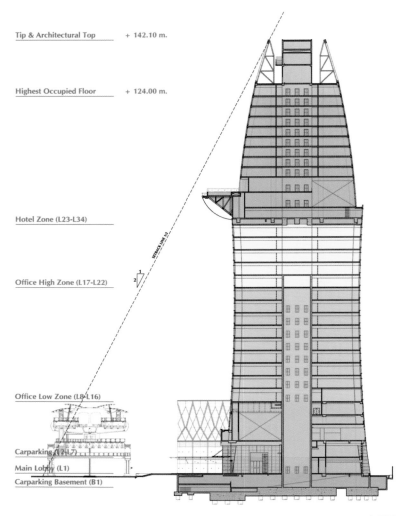

Tip & Architectural Top   + 142.10 m.

Highest Occupied Floor   + 124.00 m.

Hotel Zone (L23-L34)

Office High Zone (L17-L22)

Office Low Zone (L8-L16)

Carparking (L2-L7)

Main Lobby (L1)

Carparking Basement (B1)

Elevation　立面图

The upper portion of the sculptural tower houses a 5-star hotel with spaciously sized luxury rooms of 45 m². A striking feature that can be observed from surrounding areas is a suspended pool located at 24th floor. The swimming pool is cantilevered from the main building and provides rare unobstructed panoramic views of the city from a dynamic infinity edge.

As well as creating a sculptured environment, the Ecoplex deploys over 25% of its total building area as usable green area. Landscaped gardens are located at ground floor and additional green space is located within a rooftop area.

In addition to creating an oasis in its own right, carbon reducing technologies are employed within the building's fabric, such as Low Emission Coated laminated insulated glass fenestration, low energy consuming air conditioning systems and grey water recycling systems to name but a few.

The Park Ventures Ecoplex on Witthayu 位于曼谷市连接市区和郊区的三条主干道组成的区域网络中。这个生态型多用途建筑综合体是从曼谷中心商务区中央位置的一个类似于停车场的地块上建立起来的都市绿洲。这个位于充满活力的路口一角的建筑像是一个著名的泰国雕塑"Wai"符号致敬。建筑的南北朝向设计对于利用太阳走向进行最大限度地降低能耗战略具有非比寻常的意义。

Park Ventures 占地约 8 145 平方米，整个大楼地面上有 33 层，地下一层半是地下室。总室内建筑面积达到 81 450 平方米。包括 13 层的商业办公空间和一个拥有 242 间客房的国际水准的五星级奢华酒店。这一建筑还拥有一个可以停车 580 辆的停车场。

这个多功能综合体建筑设有少见的大型无柱办公空间。传统的中心服务核心区设计模式使建筑室内部分可以进行方正户型的布局，提高了商业办公租赁的灵活性。

这个地标性大楼高层位置设有拥有 45 平方米奢华客房的 5 星级酒店。从 24 层的悬式泳池处可以观看到附近独特迷人的市景。泳池从主建筑体上延伸出来，在这一罕见的位置可以 360 度地欣赏到充满活力的无边全景。

同样为了打造一个标志性活动空间，Ecoplex 项目中有四分之一的总建筑面积被用来作为实用性绿色区域。除了一楼的景观园林之外，建筑顶层还设置有绿地花园。

除了项目本身打造的生态景观以外，建筑外观设计上也大量使用了降低碳排放的先进技术。采用了中空复合型低排放涂层的玻璃门窗系统，低能耗暖通系统和灰水循环利用系统等各种绿色建筑技术。

Rendering　效果图

PLAN AND DESIGN FOR LONGTANG HOT SPRING, CHANGZHOU
江苏常州恐龙谷温泉酒店

Architects: Werkhart World Wide 5-Star Alliance
Location: Changzhou, Jiangsu, China
Area: 15,579 m²

设计机构：五合国际
项目地点：中国江苏省常州市
面积：15 579 平方米

This project is located in Changzhou new north district Hehai Road. Dragon Soup hot spring to hot springs as the core, set hot springs, SPA, dining, conference, accommodation for the integration of integrated project. Planning in order to "landscape around the building, construction encloses garden" layout principles. The middle road and nearby green belts around the spa reception center, food and beverage center and the hot spring of a group of buildings. At the same time, this group of public and the back of the soup accommodation area surrounded by outdoor hot spring park. Along the urban road public building with metal and glass pure house symbolizes hot rocks and water. Create a creative building modeling, embodies the "ecological, natural, fashion, elegant" creation theme.

　　本项目位于常州新北区河海东路。龙汤温泉以温泉为核心，集温泉、SPA、餐饮、会议、住宿为一体的综合项目。规划上以"景观围绕建筑，建筑围合庭院"的布局原则。沿河海东路及四周的绿化带围绕着温泉接待中心、餐饮中心和温泉会所组成的一组建筑。同时这组公建和后面的汤宿区包围着室外温泉公园。沿城市道路的公共建筑以金属和玻璃的纯净寓喻温泉的岩石和水。创造出一个富有创意的建筑造型，体现了"生态、自然、时尚、典雅"的创作主旨。

Site Plan　总平面图

图书在版编目（CIP）数据

世界建筑事务所精粹：全3册 / 深圳市博远空间文化发展有限公司编． — 天津：天津大学出版社，2013.5
　ISBN 978-7-5618-4622-3

　Ⅰ．①世… 　Ⅱ．①深… 　Ⅲ．①建筑设计－作品集－世界
Ⅳ．① TU206

中国版本图书馆 CIP 数据核字 (2013) 第 066708 号

责任编辑　郝永丽
策划编辑　刘谭春

世界建筑事务所精粹 Ⅰ 　　　　深圳市博远空间文化发展有限公司　　编

出版发行　天津大学出版社
出 版 人　杨欢
地　　址　天津市卫津路 92 号天津大学内（邮编：300072）
电　　话　发行部 022-27403647
网　　址　publish.tju.edu.cn
印　　刷　深圳市彩美印刷有限公司
经　　销　全国各地新华书店
开　　本　245 mm×330 mm
印　　张　60
字　　数　810 千
版　　次　2013 年 5 月第 1 版
印　　次　2013 年 5 月第 1 次
定　　价　998.00 元（共 3 册）

（凡购本书，如有质量问题，请向我社发行部门联系调换）